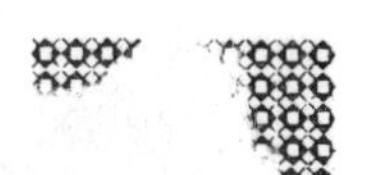

D0240584

Aliphatic nucleophilic substitution

Cambridge Chemistry Texts

Aliphatic nucleophilic substitution

S. R. HARTSHORN
Senior Demonstrator in Physical Chemistry
University of Durham

CAMBRIDGE
at the University Press 1973

Published by the Syndics of the Cambridge University Press
Bentley House, 200 Euston Road, London NW1 2DB
American Branch: 32 East 57th Street, New York, N.Y. 10022

Library of Congress Catalogue Card Number: 72–96675

ISBN: 0 521 20177 2 Hard covers
0 521 09801 7 Paperback

Printed in Great Britain by
William Clowes & Sons, Limited
London, Beccles and Colchester

Contents

Preface

The types of reactions covered by the title comprise a large and important part of organic chemistry. It is not the aim of this book to provide a comprehensive account of nucleophilic substitution reactions, but rather to describe some of the more recent work in this field, in particular that of the last decade. The approach has necessarily been selective, but it is hoped that the topics chosen for discussion will illustrate some of the more important aspects of our current understanding of the subject. Many of the established concepts have not been dealt with at great length, although the introduction should provide the beginner with the necessary background. The emphasis throughout the book has been placed on recent developments, in both the establishment of new ideas and the modification of older views. It is hoped that this small book will serve not only as an introduction to the subject, but also as a complementary text to those already available.

I should like to thank Professor V. J. Shiner, Jr, Dr G. Kohnstam, and in particular Dr R. B. Moodie for many helpful and valuable discussions. It is a pleasure to express my gratitude to Dr K. Schofield, whose advice and enthusiasm have been a particular source of help and encouragement to me. Finally, I should like to thank my many friends in the Chemistry Department of the University of Exeter, who helped to create such a pleasant atmosphere in which to write this book.

Exeter 1972 S.R.H.

1 Introduction

1.1. Aliphatic nucleophilic substitution

Nucleophilic substitution reactions involve the replacement of one functional group, X, by another, N, in such a way that N supplies a pair of electrons to form the new bond and X departs with the pair of electrons from the old bond [*1.1*].

$$N: + RX \longrightarrow RN + X: \qquad [1.1]$$

As implied by the title, this book will be limited to a discussion of examples in which the reacting centre is a saturated carbon atom. The general scheme, [*1.1*], allows some variation in the type of charge carried by N and X; initially N may be neutral or negatively charged and X may be neutral or positively charged. Examples of reactions belonging to all of the four possible charge-types are known:

N *negative*, X *neutral*. This represents the most frequently encountered charge-type; examples include the Finkelstein reaction [*1.2*], the Williamson ether synthesis [*1.3*], and alkaline hydrolysis reactions [*1.4*],

$$I^- + RBr \longrightarrow RI + Br^- \qquad [1.2]$$

$$RO^- + R'Br \longrightarrow ROR' + Br^- \qquad [1.3]$$

$$HO^- + RX \longrightarrow ROH + X^- \qquad [1.4]$$

N *neutral*, X *neutral*. The Menschutkin reaction [*1.5*] illustrates this charge-type.

$$R_3N + R'X \longrightarrow R_3\overset{+}{N}R' + X^- \qquad [1.5]$$

N *negative*, X *positive*. The Hofmann degradation [*1.6*] and related reactions such as [*1.7*] are examples of this charge-type.

$$HO^- + R_3\overset{+}{N}R \longrightarrow ROH + R_3N \qquad [1.6]$$
$$Cl^- + R_3\overset{+}{N}R \longrightarrow RCl + R_3N \qquad [1.7]$$

N *neutral*, X *positive*. The final charge-type is illustrated by various 'onium ion exchange reactions such as [*1.8*] and [*1.9*].

$$Me_3N + Me\overset{+}{S}Me_2 \longrightarrow Me_3\overset{+}{N}Me + SMe_2 \qquad [1.8]$$
$$Me_3N + Me\overset{+}{N}Me(C_6H_5)_2 \longrightarrow Me_3\overset{+}{N}Me + NMe(C_6H_5)_2 \qquad [1.9]$$

This large class of reactions therefore includes many of the well-known and useful reactions of organic chemistry. Nucleophilic substitutions may also involve unstable species, the substitution being an intermediate step in a more complex reaction sequence. In many cases the solvent acts as the nucleophile, and such reactions are referred to as solvolytic displacement, or solvolysis reactions.

1.2. The mechanistic classification of nucleophilic substitutions

During the early 1930s Hughes and Ingold and their co-workers recognised that the various nucleophilic substitution reactions were related by common mechanistic patterns. They rationalised many of the observations by postulating the existence of two fundamental mechanisms for such reactions (Ingold, 1969). The mechanistic classification provided the guidelines for the subsequent systematic study of this class of reactions.

1.2.1. The bimolecular mechanism, S_N2. In this one-step process the bond between the nucleophile and the reacting carbon atom is formed at the same time as the bond between the carbon atom and the leaving group is broken [*1.10*]. The nucleophile attacks the carbon atom on the

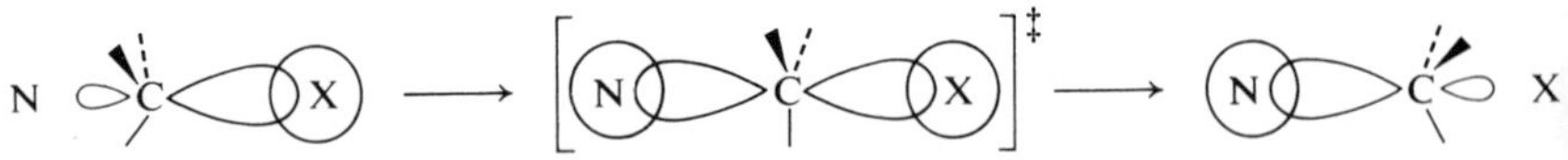

$$N \;\; C \;\; X \longrightarrow [N \cdots C \cdots X]^{\ddagger} \longrightarrow N \;\; C \;\; X \qquad [1.10]$$

side opposite to that from which X departs, leading to an inversion of configuration about the reacting carbon atom (§3.3). In the transition state N and X are attached to the reacting carbon atom by partial covalent bonds, which may be represented by the interactions of N and X with a p-orbital of the carbon atom (see p 68). This latter orbital is formed by a rehybridisation of the sp^3-orbitals of the saturated carbon atom of the substrate. The partial bonds are usually represented by dotted lines; thus the transition state may be represented by (1). Since

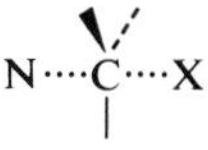

N····C····X

(1)

two molecules are simultaneously undergoing a covalency change in the rate-limiting step, the reaction is bimolecular; the mechanism is called S_N2, to signify *s*ubstitution, *n*ucleophilic, *bi*molecular (cf. §3.1).

1.2.2. The unimolecular mechanism, S_N1. The second mechanism of nucleophilic substitution involves two steps; a slow rate-limiting ionisation of RX, followed by a fast co-ordination between the intermediate thus produced and the nucleophile N [*1.11*]. The intermediate

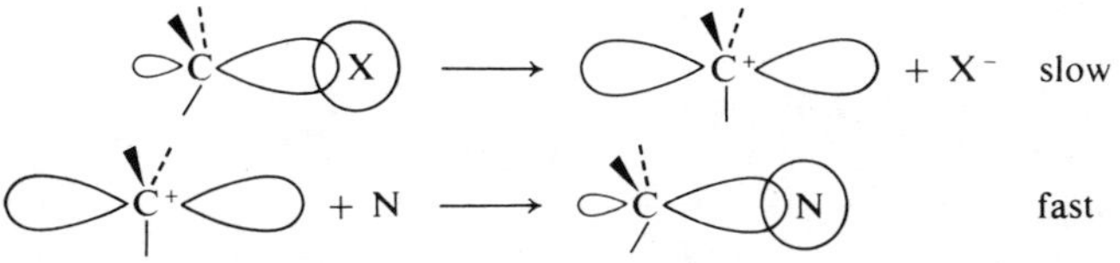

C–X ⟶ C^+ + X^- slow

C^+ + N ⟶ C–N fast [*1.11*]

is written here as it was originally proposed, as a 'free' carbonium ion, but it is now recognised that ionisation need not lead directly to a fully dissociated species. Thus the intermediate with which N co-ordinates might be an ion pair (§4.1), formed after ionisation of RX, but before dissociation to the carbonium ion R^+ free from interaction with the counter ion X^-. Since only RX is undergoing a covalency change in the rate-limiting step the reaction is unimolecular; the mechanism is written S_N1, *s*ubstitution, *n*ucleophilic, *uni*molecular (cf. §3.1).

When it was first proposed, the S_N1 mechanism encountered considerable opposition on the grounds that the ionisation of the substrate would be energetically improbable. It was argued, however, that the energy of the initial heterolysis could be compensated by the energy of solvation of the ions produced, thus the interactions between the ionic species and the solvent were seen to be of primary importance. If these interactions are assumed to be electrostatic, then there is no need to include any solvent molecules in the description of the reaction mechanism, since we are concerned only with the number of molecules which necessarily undergo a covalency change. This approach leads to the useful simplification that although an unknown number of solvent molecules are involved, they need not be considered explicitly for the purposes of defining the mechanism. It has been pointed out (Ingold, 1969) that if the solvent molecules are included then nearly all reactions in solution become multimolecular.

1.2.3. Alternative classifications. A different approach to the problem of classification was suggested by Winstein and his co-workers (Winstein, Grunwald and Jones, 1951). They called reactions nucleophilic (N) if there was a covalent interaction between the nucleophile and the substrate in the transition state of the rate-limiting step, and limiting (Lim) if there was no such interaction. The transition state was considered to be a mesomeric form of the canonical structures (2), (3) and (4), the covalent interaction being represented by giving weight to structure (4). Whilst the definitions of the mechanisms N and Lim are

$$\underset{\textbf{(2)}}{\mathrm{N\!:\;\; R-X}} \qquad \underset{\textbf{(3)}}{\mathrm{N\!:\;\; \overset{+}{R}\;\; :X}} \qquad \underset{\textbf{(4)}}{\mathrm{N-R\;\; :X}}$$

very similar to those of the mechanisms S_N2 and S_N1 respectively, the two classifications differ in one important respect. It is that S_N2 and S_N1 are regarded as being fundamentally distinct mechanisms, whereas N and Lim represent mechanisms which may be gradually merged, by adjusting the weights given to the various canonical structures. This difference becomes particularly important when the mechanisms of reactions in the borderline region are discussed (§1.3).

The idea of gradually merging mechanisms for S_N reactions was also developed by Doering and Zeiss (1953), and by Streitwieser (1962). In this approach, usually called the 'structural hypothesis', explicit account is taken of the solvation of the carbonium ion by the solvent, or other available nucleophile, and by the leaving group. It is assumed

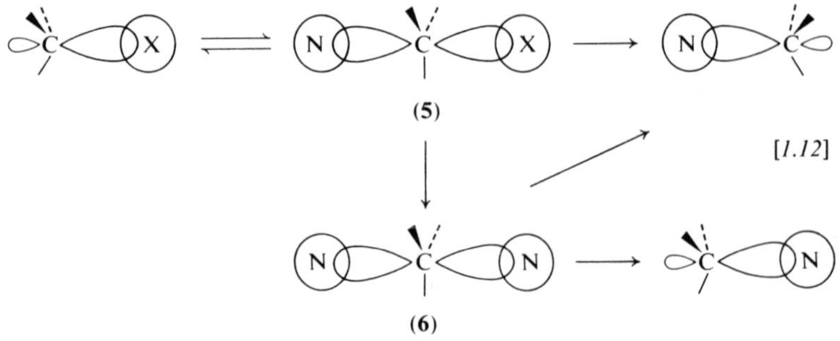

(5)

[*1.12*]

(6)

that only two solvation sites need to be considered, so that a typical nucleophilic substitution can be represented by [*1.12*].

The rate-limiting step is the formation of the intermediate (5), in which the nucleophile and the leaving group provide the solvation; each forms a partial covalent bond with a lobe of the p-orbital of the reacting carbon atom. This intermediate may collapse to give starting material with retention of configuration, or product with inversion of configuration. Alternatively, X may be replaced by another molecule of the nucleophile to give the symmetrically solvated intermediate (6), which gives rise to racemic product. The reaction pathway actually followed will depend upon the stability of the intermediate (5). The various possibilities are illustrated by the potential energy diagram in fig. 1.1.

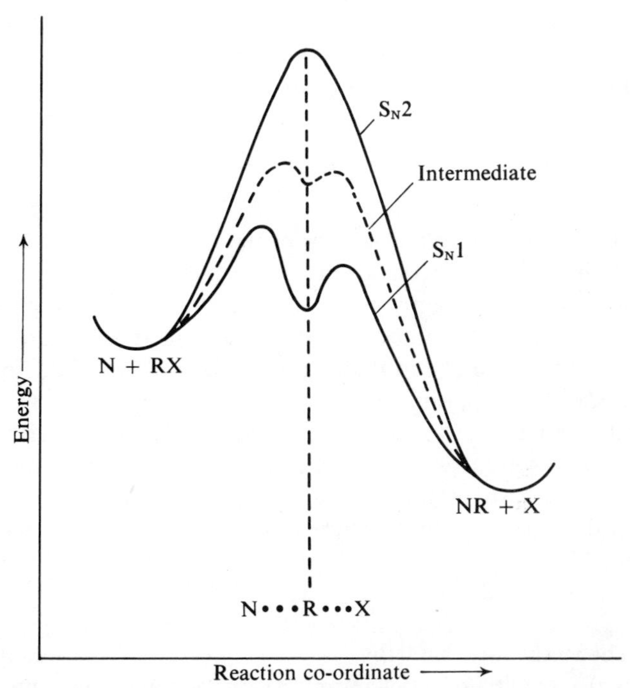

Fig. 1.1 Potential energy curves for different nucleophilic substitutions.

For a given nucleophile and a given leaving group, the stability of (5) will parallel the stability of the hypothetical carbonium ion obtained by ionising RX. In extreme cases this ionisation would be energetically so unfavourable that (5) would only be formed with nucleophilic assistance. In such cases (5) is best thought of as a transition state for a process which closely resembles the S_N2 mechanism. As (5) becomes more stable it is possible to regard it as a true reaction intermediate, a circumstance marked by the appearance of a shallow minimum in the potential energy curve. The transition state leading to (5) is now more easily formed and requires less nucleophilic assistance. The limiting case corresponds to the situation in which there is no interaction between the nucleophile and the reacting carbon atom in the transition state leading to (5), and it thus closely resembles the S_N1 mechanism. In terms of the structural hypothesis, therefore, the types of behaviour described by the mechanisms S_N1 and S_N2 are regarded as extremes of a single mechanism.

A further mechanism for S_N reactions that is sometimes discussed is the 'push–pull' mechanism of Swain (1948). In this, a nucleophile attacks the carbon atom at the reaction centre while at the same time a solvent molecule, or some other electrophile, assists the departure of the leaving group (7); the mechanism is thus termolecular. It was originally

$$[N\cdots R\cdots X\cdots S]^{\ddagger}$$

(7)

proposed to rationalise the kinetics of the reaction between triphenylmethyl chloride and methanol in benzene. However, the reaction in question was subsequently found not to show a firm second order dependence on methanol, but to have instead an apparent order which rose smoothly from one to three as the concentration of the methanol increased. There appears to be no necessity therefore to attach any general significance to a termolecular mechanism. Furthermore, benzene is a solvent with properties very different from those of the polar solvents usually used for S_N reactions, so that any mechanism proposed for a reaction in benzene may not be applicable to other solvents.

1.3. The mechanistic borderline

When the duality of mechanism for nucleophilic substitutions was proposed, S_N1 and S_N2 were regarded as being extremes of a graded range of mechanisms. The possible existence of borderline cases (i.e. reactions showing characteristics of both the S_N1 and S_N2 mechanisms) was recognised, and these were thought of as having intermediate mechanisms. However, Hughes and Ingold and their co-workers have never described these intermediate mechanisms in detail, and have indeed often treated them simply as mixtures of S_N1 and S_N2 processes. In the latest discussion (Ingold, 1969) it is emphasised that S_N1 and S_N2 are fundamentally distinct mechanisms, but that we should not expect the change from a bimolecular to a unimolecular mechanism to be accompanied necessarily by a sharp change in the characteristics of the reactions.

A more detailed account has been presented by Gold (1956). In this the role played by the nucleophile is seen to diminish in importance as we go from conditions typical for an exclusively S_N2 reaction towards the borderline region. In this progression the breaking of the C–X bond [*1.10*] gradually assumes greater importance, and the formation of the new C–N bond less importance, in contributing to the free energy change for the formation of the transition state. The reaction is bimolecular, i.e. S_N2, over the whole range, although the transition state changes from a 'tight' structure, with a strong C–N interaction, at the extreme, to a 'loose' structure, with a weak C–N interaction, in the borderline region (cf. §2.1.2). Since, in the borderline region, the interaction between the nucleophile and the reacting carbon atom becomes of secondary importance in determining the free energy change accompanying the formation of the transition state, it seems plausible to attribute comparable importance to a range of transition state structures having varying C–N interactions, and including the case in which there is no such interaction. Thus in the progression from the S_N2 extreme through the borderline region to the S_N1 extreme, there is both a gradual change in the bimolecular reaction, from a tight to a loose transition state, and a gradual increase in the contribution from the unimolecular reaction. In this sense the reaction may be described as a mixture of concurrent bimolecular and unimolecular processes.

This view of the mechanism in the borderline region allows considerable variation in the characteristics of nucleophilic substitutions, but at the same time it retains the idea that there is an abrupt difference between the S_N1 and S_N2 mechanisms as far as molecularity is concerned, i.e. that they are fundamentally distinct mechanisms. As Gold points out this is a consequence of the quantisation of matter, since it is not possible to have a gradual transition between no molecule and one molecule.

Some support for the view that the nature of the transition state may vary in the borderline region comes from a study of the reactions of 4-phenoxybenzyl chloride (8*a*) and 4-methoxybenzyl chloride (8*b*) with various nucleophiles in 70 per cent aqueous acetone.† In this solvent (8*a*) and (8*b*) undergo unimolecular hydrolysis, but both compounds react with added nucleophiles by a bimolecular mechanism (cf. §4.3.4).

$RO-C_6H_4-CH_2-Cl$

(8)

a R = C_6H_5
b R = CH_3

The second-order rate constants for these reactions with added nucleophiles are shown in table 1.1 (Kohnstam, Queen and Ribar, 1962). The

TABLE 1.1 *Second-order rate constants for the bimolecular reactions of 8a and 8b with added nucleophiles in 70 per cent aqueous acetone at 20 °C*[a]

Nucleophile	$10^5 k_2$ (l mol^{-1}s^{-1})		$k_2(8a)/k_2(8b)$
	8*a*	8*b*	
$C_6H_5.SO_3^-$	20	0.16	125
NO_3^-	32	0.23	139
F^-	42	0.49	86
Cl^-	66	0.76	87
Br^-	79	5.09	15.5
N_3^-	345	71.03	4.9

[a] Data from Kohnstam, Queen and Ribar (1962).

value of k_2 increases with increasing strength of the nucleophile for both (8*a*) and (8*b*). Of particular interest, however, is the last column which shows the relative rates of reaction of (8*a*) and (8*b*) with the different nucleophiles. With the weak nucleophiles, $C_6H_5.SO_3^-$ and NO_3^-, the ratio $k_2(8a)/k_2(8b)$ has the same value as the ratio of first-order rate constants for the unimolecular hydrolysis ($k_1(8a)/k_1(8b)$ = 135 in the absence of added electrolyte). The value of $k_2(8a)/k_2(8b)$ decreases as the strength of the attacking nucleophile increases, being 4.9 for azide ion.

These observations were interpreted in the following manner. When the operation of the S_N1 mechanism is not energetically unfavourable relative to the S_N2 mechanism, as in the present systems, the S_N2 attack

† 70 per cent aqueous acetone refers to a solvent made by mixing together 70 parts by volume of acetone and 30 parts by volume of water. Volume percentages are traditionally used, although a composition so defined is a function of the temperature.

by the nucleophile does not commence until some heterolysis of the C–Cl bond has occurred. The weaker the nucleophile, the later the stage in the heterolysis at which this attack begins, until with the weakest nucleophile studied, water, attack occurs only after the heterolysis is complete and the reaction is S_N1. The observed relative reactivities of (8*a*) and (8*b*) towards the different nucleophiles indicate that the nature of the S_N2 transition state varies with the strength of the attacking nucleophile. A strong nucleophile, such as azide ion, has a tight transition state, probably close to the extreme structure for an S_N2 reaction, and a low relative reactivity is observed. On the other hand, weak nucleophiles, such as benzenesulphonate or nitrate ion, have loose transition states with some carbonium ion character; the relative reactivity is now close to that observed in an S_N1 reaction. It should be emphasised that all reactions with the nucleophiles are bimolecular, in spite of the variation in the structure of the transition states. It was, however, pointed out that the results do not exclude the possibility of the nucleophiles reacting, in part, with a fully developed carbonium ion by a concurrent S_N1 mechanism.

Although data are available for many borderline reactions the interpretation of the results is often made uncertain by the assumptions and corrections that have to be applied when treating the experimental observations. For example, in reactions involving added nucleophiles corrections have to be applied for salt effects and for the incomplete dissociation of the salt used to supply the nucleophile. An experiment designed to minimise these uncertainties was the exchange between isotopically labelled thiocyanate ion and *para*-substituted benzhydryl-thiocyanates in acetonitrile [*1.13*] (Ceccon, Papa and Fava, 1966).

$$X-C_6H_4-CH(SCN)-C_6H_4-X' + {}^*SCN^- \rightleftharpoons X-C_6H_4-CH({}^*SCN)-C_6H_4-X' + SCN^- \qquad [1.13]$$

Since the concentration of the nucleophile is constant throughout any given experiment there can be no difficulty arising from the mass-law effect (§3.2.1). By using a high concentration of an inert electrolyte ($NaClO_4$), having the same cation as the ionic reagent (Na*SCN), it is possible to swamp the salt effects and at the same time to keep constant the ion-pair dissociation fraction of the ionic reagent. It was found that under these conditions the observed first-order rate constant for the initial part of the exchange, k_{obs}, could be expressed as the sum of a first-order process, with rate constant k_1, and a second-order process, with rate constant k_2, (1.1). The term [NaSCN] represents the stoichio-

$$k_{obs} = k_1 + k_2 \text{[NaSCN]} \quad (1.1)$$

metric concentration of sodium thiocyanate. Values of k_1 and k_2 for exchange reactions with different substituents X and X′ (table 1.2) were calculated from the intercepts and slopes respectively of graphs of k_{obs} against [NaSCN] (fig. 1.2).

TABLE 1.2 *First- and second-order rate constants for isotopic exchange between benzhydryl thiocyanates and* NaSCN [*1.13*], *in acetonitrile at 70 °C*[a]

X	X′	$10^6 k_1 (s^{-1})$	$10^4 k_2 (l\ mol^{-1} s^{-1})$
NO_2	NO_2	–	3.02
NO_2	H	–	2.47
Cl	H	4.66	5.80
H	H	7.11	8.56
CH_3	H	410[b]	–
CH_3	CH_3	7300[b]	–

[a] Data from Ceccon, Papa and Fava (1966); $[NaSCN] + [NaClO_4] = 0.1\ mol\ l^{-1}$.
[b] Extrapolated from data at other temperatures.

The exchange takes place entirely by a bimolecular mechanism when electron withdrawing substituents are present, and entirely by a unimolecular mechanism when electron donating substituents are present. In the intermediate cases the exchange takes place by concurrent unimolecular and bimolecular mechanisms. Thus the gradual transition from one extreme of mechanism to another has been achieved by relatively minor variations in the structure of the reacting substrate. The effects of these changes in structure on the magnitudes of k_1 and k_2 are typical of those observed on the rates of S_N1 and S_N2 reactions respectively. In addition, a more careful analysis of the effect of the structural changes on the values of k_2 (§3.6.1) shows evidence for a variation in the nature of the transition state involved in the bimolecular reaction. The results therefore support the idea that nucleophilic substitutions in the mechanistic borderline can be considered in terms of concurrent S_N1 and S_N2 mechanisms.

Many examples of borderline behaviour arise in solvolytic displacements. In such cases it is not possible to gain any information about the molecularity of the rate-limiting step from the observed kinetic order, because the solvent is the substituting agent and it is always present in large excess. The nature of the transition state of the rate-limiting step

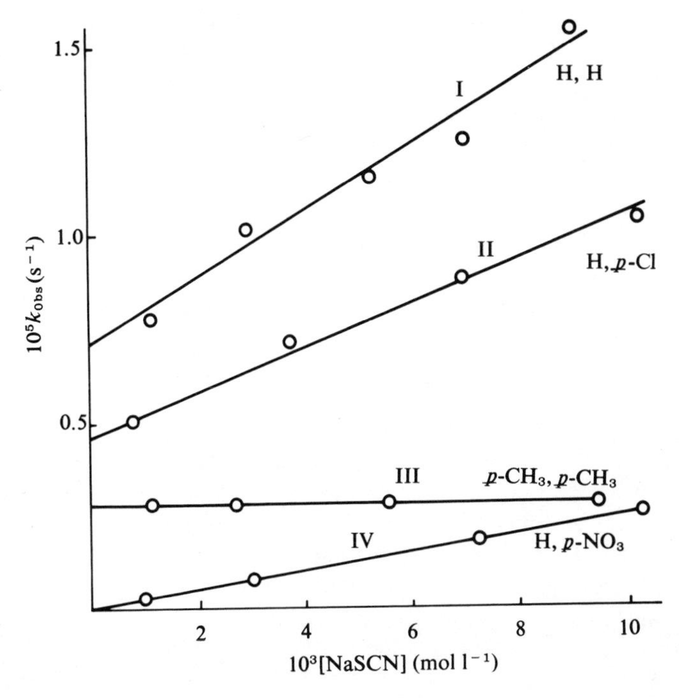

Fig. 1.2 The dependence of the rate of exchange, k_{obs}, on the concentration [NaSCN]: I, II and IV at 70 °C; III at 0 °C.

must be inferred from other properties of the reaction; the characterisation of mechanism forms the subject of chapter 3. In spite of the number of probes available, the description of the mechanisms of borderline reactions remains a difficult and controversial problem. It is still not possible to decide whether borderline reactions proceed by concurrent S_N1 and S_N2 mechanisms or by a single intermediate mechanism. The difficulty is that any mechanistic investigation relies on a comparison being made between the experimentally observed properties and those calculated for a model process. Such an approach usually involves making one or more assumptions which may render the comparison nugatory.

The lack of success in dealing with the mechanistic borderline probably reflects a limitation on the kind of information that we can hope to procure from a mechanistic study. Any property derived from kinetic measurements is an average property of a large number of molecular acts. It is usual to assume that such a property applies to every molecular act of the reaction. However, it follows from the uncertainty in the extent to which the properties of a reacting system may be specified that there will be, in fact, many configurations of reacting molecules, of slightly different energies, occurring over different lengths of time. In other words, we can never hope to describe in detail every molecular act of a reaction. We cannot therefore expect to distinguish between different reaction pathways, which for a given reaction involve transition states of very similar energies. This appears to be the situation in the mechanistic borderline for nucleophilic substitutions, so that it is important to realise that no matter what interpretation we choose for the mechanism, we can never be certain that it is unique.

1.4. A further classification

We have seen in an earlier section that nucleophilic substitutions may be classified according to the molecularity of the rate-limiting step, and that this requires only two fundamental mechanisms, S_N1 and S_N2, to be postulated. In this section we shall describe some examples which are frequently referred to by other symbols. This further classification is not an alternative to that already described, but instead it allows some special cases to be recognised.

1.4.1. The $S_N2(C^+)$ mechanism. The S_N1 mechanism [*1.11*] is characterised by a slow heterolysis of the reacting substrate followed by a rapid co-ordination of the intermediate formed with a nucleophile. If the rate of the heterolysis becomes comparable to, or greater than, that of the co-ordination, then the mechanism involves a bimolecular reaction between the nucleophile and the intermediate in the rate-limiting step. This type of mechanism is called $S_N2(C^+)$, to signify the attack of the nucleophile on a preformed (carbonium ion) intermediate. This mechanism requires conditions favourable for the formation of stable carbonium ions. An example is the reaction of triphenylmethyl chloride with various hydroxylic reagents in nitromethane [*1.14*]. The first equilibrium is established very rapidly. The second equilibrium

$$(C_6H_5)_3CCl \rightleftharpoons (C_6H_5)_3C^+ + Cl^- \quad \text{fast}$$
$$(C_6H_5)_3C^+ + ROH \rightleftharpoons (C_6H_5)_3COR + H^+ \quad \text{slow} \qquad [1.14]$$

may be displaced to the product side by the addition of pyridine, which removes the hydrogen chloride formed; there is some evidence to suggest that the pyridine also forms a complex with the carbonium ion. Under these conditions second-order kinetics are observed, first order in the concentration of the substrate and first order in the concentration of the hydroxylic reagent. The reactivity of the latter species increases with its nucleophilic strength, the relative rates for the series EtOH : H_2O : C_6H_5OH being 20:5:1 (cf. §1.5.3). The observed order of reactivity and the second-order kinetics are both consistent with the $S_N2(C^+)$ mechanism being operative.

1.4.2. The S_N2' mechanism. Many bimolecular displacements involving allylic systems are found to give rearranged products. Thus the reaction of α-methylallyl bromide (9) with radio-bromide ion in acetone gives a mixture of isomeric bromides (10) and (11). The major product (10) is obtained by the normal S_N2 mechanism, but the formation of the rearranged product (11) is an example of the S_N2' mechanism; bi-

$$\underset{(9)}{CH_2{=}CH{-}CH(CH_3){-}Br} \xrightarrow{S_N2} \underset{(10)}{CH_2{=}CH{-}CH(CH_3){-}{}^*Br}$$

$$\xrightarrow{S_N2'} \underset{(11)}{{}^*Br{-}CH_2{-}CH{=}CH{-}CH_3}$$

molecular nucleophilic substitution with concurrent rearrangement [1.15].

$$N\!: \; >C{=}C{-}C{-}X \longrightarrow N{-}C{-}C{=}C< \quad X\!: \qquad [1.15]$$

The S_N2' mechanism is seemingly easy to characterise, but it has proved to be very difficult to distinguish it from other forms of substitution accompanied by rearrangement. Indeed, so few unambiguous examples can be cited that some doubt has been expressed about its existence. The reaction of α-methylallyl chloride with diethylamine to give crotyldiethylamine, for example, was originally described as an example of the S_N2' mechanism [1.16]. However, it has been pointed out

$$Et_2\ddot{N}(H)\text{—}CH_2\text{=}CH\text{—}CH(CH_3)\text{—}Cl \longrightarrow \{Et_2N(H)\text{—}CH_2\text{—}CH\text{=}CH(CH_3)\}^+ Cl^- \longrightarrow Et_2N\text{—}CH_2\text{—}CH\text{=}CH(CH_3) + HCl \quad [1.16]$$

that since the hydrogen atom of the amine is apparently essential to the reaction, the rearrangement might proceed through the hydrogen bonded structure (12), and might therefore be an example of the S_Ni'

(12)

mechanism (Ingold, 1969) (see p. 19). The observed second-order kinetics are consistent with both interpretations, since the kinetic form indicates only that both molecules are present in the transition state, and gives no information about their mode of interaction.

The stereochemistry of the S_N2' mechanism is of interest because it was shown that in the reaction of piperidine with the cyclohexenyl compound (13) the incoming nucleophile and the departing leaving group were both *trans* to the methyl group, and therefore on the same side of the substrate [*1.17*] (Stork and White, 1956). This situation is to

(13) [*1.17*]

be contrasted with that of the normal S_N2 mechanism, in which the incoming nucleophile and the leaving group are on opposite sides of the substrate molecule. Although this result suggests that the S_N2' mechanism is a synfacial process, the possibility cannot be excluded that the reaction in question occurs by the S_Ni' mechanism involving the hydrogen bonded intermediate (14).

(14)

The predictions of simple MO theory are of interest in this connection. For a concerted process of the above type in which bond breaking runs ahead of bond making, the stereochemistry should be determined by the symmetry of the lowest vacant MO of the allylic cation (15). The transition state (16) should therefore involve the allylic system in a

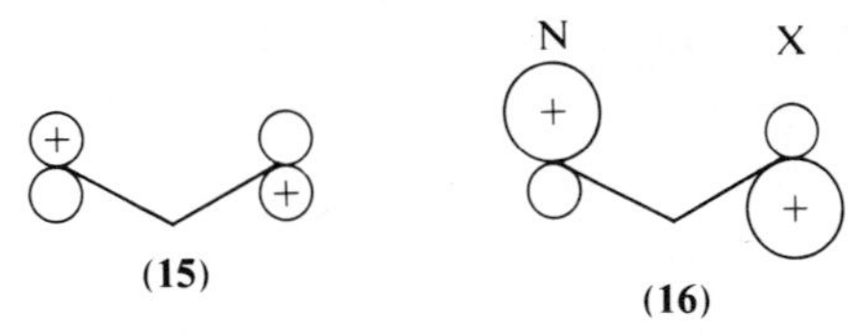

(15) **(16)**

bonding interaction with the nucleophile (N) and an antibonding interaction with the leaving group (X), which leads to the prediction that the process should be synfacial (Anh, 1968). Further MO considerations seem to indicate that if N and X can have a suitable orbital interaction then the quasi-cyclic transition state that results will be energetically more favourable than the arrangement in which N and X do not interact. This result would imply that the S_Ni' rather than the S_N2' mechanism is a more appropriate description of these types of reactions (Jefford, Sweeney, Hill and Delay, 1971).

1.4.3. The S_N1' mechanism. The rearranged products observed in reactions of allylic systems are sometimes formed by a unimolecular mechanism involving a carbonium-ion (or ion-pair) intermediate [*1.18*]. The mesomeric carbonium ion can react at either the α- or the γ-carbon atom; reaction at the latter position leads to rearranged product by the S_N1' mechanism. Before attributing rearrangement in the products to the operation of the S_N1' mechanism, it is necessary to exclude the possibility that the rearrangement is caused by isomerisation of the starting material followed by substitution by the normal S_N1 mechanism.

$$\gt C{=}C{-}C(R){-}X \longrightarrow \left[\gt C{\cdots}C{\cdots}C\lt^{R}\right]^{+} \begin{cases} \xrightarrow{S_N1} \gt C{=}C{-}C(R){-}N \\ \xrightarrow{S_N1'} N{-}C{-}C{=}C\lt^{R} \end{cases} \qquad [1.18]$$

Rearrangements accompanying unimolecular reactions are not restricted to unsaturated systems, but are also observed with many saturated substrates. There is good evidence to suggest that the solvolysis of neopentyl bromide (17) in aqueous ethanol occurs by a unimolecular mechanism. The products of solvolysis, t-amyl alcohol (18), t-amyl ethyl ether (19), and trimethylethylene (20), all have rearranged carbon skeletons (Dostrovsky and Hughes, 1946). It is assumed that the initial ionisation gives the neopentyl cation which then rearranges to the more stable t-amyl cation [*1.19*]. Since non-rearranged products

$$\underset{\mathbf{(17)}}{Me{-}C(Me)_2{-}CH_2Br} \longrightarrow Me{-}C(Me)_2{-}\overset{+}{C}H_2 \longrightarrow \qquad [1.19]$$

$$Me_2\overset{+}{C}{-}CH_2Me \longrightarrow \begin{cases} \underset{\mathbf{(18)}}{Me{-}C(Me)(OH){-}CH_2Me} \\ \underset{\mathbf{(19)}}{Me{-}C(Me)(OEt){-}CH_2Me} \\ \underset{\mathbf{(20)}}{Me{-}C(Me){=}CHMe} \end{cases}$$

are apparently not formed, the rearrangement occurs either simultaneously with the heterolysis of the C–Br bond or else very soon afterwards. The timing of the rearrangement can have important consequences, because the migration of the methyl group with its bonding electrons might assist the bond heterolysis, and therefore lead to an abnormally high rate of reaction (§5.2.2).

Another rearrangement frequently observed in the unimolecular reactions of saturated systems is the hydride shift [*1.20*]. It was shown

$$-\overset{+}{C}H_2.CH_2.\overset{+}{C}H.CH_2- \begin{array}{l} \nearrow -CH_2.\overset{+}{C}H.CH_2.CH_2- \\ \searrow -\overset{+}{C}H.CH_2.CH_2.CH_2- \end{array} \qquad [1.20]$$

by careful analysis that compounds arising by hydride shifts are invariably found amongst the products in systems thought to react by a carbonium ion-like intermediate (Whiting, 1966). It has been suggested, moreover, that the observation of hydride shifts might be a useful diagnostic test for the involvement of carbonium ion (or ion-pair) intermediates in nucleophilic substitutions.

1.4.4. The S_Ni mechanism. A common method for preparing alkyl chlorides is the reaction between an alcohol and thionyl chloride [*1.21*].

$$ROH + SOCl_2 \longrightarrow \underset{(\mathbf{21})}{ROSOCl} \longrightarrow RCl + SO_2 + HCl \qquad [1.21]$$

It is known that the reaction proceeds through the intermediate formation of an alkyl chlorosulphite (21). The chloride produced often has the same configuration as that of the original alcohol, and in order to explain this retention of configuration it was proposed that the chlorosulphite intermediate undergoes an intramolecular nucleophilic substitution, S_Ni (22), in which the chlorine atom effects an internal dis-

O
C S=O
Cl

(22)

placement with loss of sulphur dioxide (Cowdrey, Hughes, Ingold, Masterman and Scott, 1937).

The reactions for which the S_Ni mechanism was proposed exhibit many of the characteristics of the S_N1 mechanism, and, in order to emphasise these similarities Cram (1953) proposed an ion-pair mechanism [*1.22*] to describe the decomposition of the chlorosulphite intermediate. The initial ionisation to form the ion pair (R^+OSOCl^-) would be the rate-limiting step; rapid loss of sulphur dioxide from the anion then leads to the new ion pair (R^+Cl^-), and recombination of this ion pair gives the alkyl chloride with the same configuration as that of the original alcohol.

Although the ion-pair mechanism [*1.22*] and the S_Ni mechanism are alternative rationalisations of the same experimental observations, the

$$\begin{array}{ccc} ROSOCl & \longrightarrow & (R^+OSOCl^-) \\ & & \downarrow \\ RCl & \longleftarrow & (R^+Cl^-) + SO_2 \end{array} \qquad [1.22]$$

former should not be used as an alternative description of the S_Ni mechanism, which has been defined explicitly in terms of the cyclic internal displacement (22). It seems desirable to maintain this distinction for another reason; for, whereas the S_Ni mechanism will always lead to retention of configuration, the ion-pair mechanism need not. Racemisation is possible at the ion-pair stage before internal return to give product.

This point is illustrated by a recent study of the decomposition of aralkyl thiocarbonates in inert solvents at elevated temperatures [*1.23*] (Kice, Scriven, Koubek and Barnes, 1970). The dependence of the rate

$$\underset{\displaystyle C_6H_5}{\overset{}{Ar.\underset{|}{C}H.O.\overset{\displaystyle O}{\overset{\|}{C}}.SR}} \longrightarrow \underset{\displaystyle C_6H_5}{Ar.\underset{|}{C}H.SR} + CO_2 \qquad [1.23]$$

of decomposition on the ionising power of the solvent and on the structure of the aralkyl group indicates that a polar transition state is involved. The variation of the rate of decomposition with the structure of the thioalkyl group suggests, in addition, that the carbon–sulphur bond is partially broken in the transition state of the rate-limiting step. In benzonitrile, optically active *p*-chlorobenzhydryl *S*-methyl thiocarbonate (23) was found to racemise several times faster than it decomposed to the sulphide and carbon dioxide; the following mechanism [*1.24*] was proposed. The first formed ion-pair (24) returns to starting material faster than it decomposes to the second ion pair (25). The internal return of (24) occurs with partial loss of optical activity and therefore accounts for the fact that (23) racemises faster than it decomposes. The authors call the mechanism [*1.24*] S_Ni, although no cyclic intermediate is involved, and the sulphide (26), formed by internal return of the second ion pair (25), was found to be racemic.

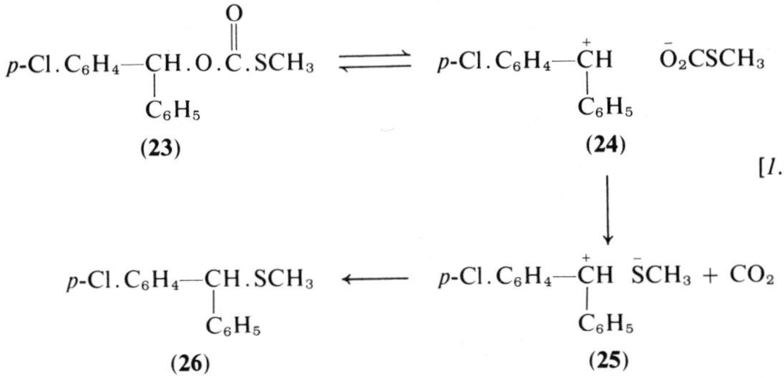

$$p\text{-Cl.C}_6\text{H}_4\text{—}\underset{\underset{\mathbf{(23)}}{\text{C}_6\text{H}_5}}{\text{CH}}\text{.O.}\overset{\text{O}}{\overset{\|}{\text{C}}}\text{.SCH}_3 \rightleftharpoons p\text{-Cl.C}_6\text{H}_4\text{—}\underset{\underset{\mathbf{(24)}}{\text{C}_6\text{H}_5}}{\overset{+}{\text{C}}\text{H}}\quad \overset{-}{\text{O}}_2\text{CSCH}_3 \qquad [1.24]$$

$$\downarrow$$

$$p\text{-Cl.C}_6\text{H}_4\text{—}\underset{\underset{\mathbf{(26)}}{\text{C}_6\text{H}_5}}{\text{CH}}\text{.SCH}_3 \longleftarrow p\text{-Cl.C}_6\text{H}_4\text{—}\underset{\underset{\mathbf{(25)}}{\text{C}_6\text{H}_5}}{\overset{+}{\text{C}}\text{H}}\ \ \overset{-}{\text{S}}\text{CH}_3 + \text{CO}_2$$

1.4.5. The S_Ni' mechanism. An additional type of internal displacement is encountered in the reactions of thionyl chloride with allylic alcohols [*1.25*]. In such cases the decomposition of the chlorosulphite

$$\text{RCH=CH—CH}_2\text{OH} \xrightarrow{\text{SOCl}_2} \text{RCH=CH—CH}_2\text{OSOCl} \xrightarrow{-\text{SO}_2} \text{R}\overset{\text{Cl}}{\overset{|}{\text{C}}}\text{H—CH=CH}_2 \qquad [1.25]$$

intermediate occurs with allylic rearrangement, which may be represented by the concerted cyclic mechanism S_Ni' (27). Another possibility is that the decomposition occurs by an ion-pair mechanism of which (28) is the first formed intermediate. The preference of allylic

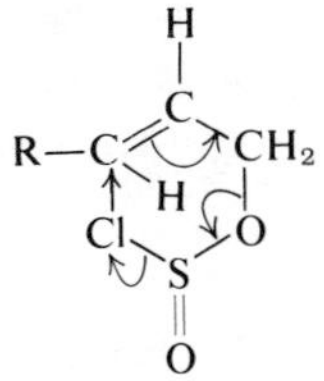

(27) 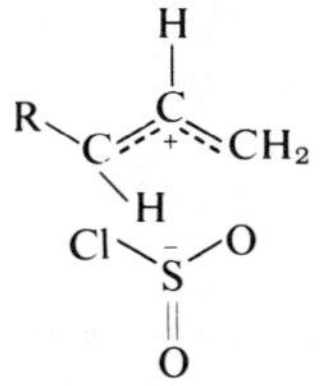 (28)

chlorosulphites for decomposing with rearrangement (Wolfe and Young, 1956), suggests that if the ion-pair representation is used it is necessary to postulate a rigidly orientated structure, so that internal return of chloride, after loss of sulphur dioxide, gives predominantly the rearranged product.

The observations made on the rearrangements of allylic thiobenzoates are probably relevant in this connection (Smith, 1961). It was found that the rate of the rearrangement [*1.26*] was very insensitive to the ionising power of the solvent. This indicates a mechanism with very

[*1.26*]

little development of charge in the transition state. The similarity between this rearrangement and the decomposition of allylic chlorosulphites would suggest that the latter reaction is better represented by the cyclic mechanism (27) than by the ion-pair mechanism involving (28). This should not, however, be taken as evidence against the operation of an ion-pair mechanism as a possible alternative for the S_Ni mechanism in general. The allylic system might be a special case in that the structure is a very favourable one for the operation of the concerted mechanism.

1.5. Catalysed nucleophilic substitutions

We have already seen (§1.2.3) that the 'push-pull' mechanism of Swain emphasises the importance of an electrophile in assisting the departure of the leaving group. Whilst in general it is not necessary to postulate such a specific interaction, in the special case of catalysed nucleophilic substitutions just such a process may assume importance. Many reactions of alkyl halides and alkyl arenesulphonates, in a variety of solvents, are found to be catalysed by Lewis acids, the catalysis being homogeneous or heterogeneous depending upon the conditions. In aprotic solvents catalysis by protonic acids and even hydroxylic reagents is also frequently observed.

1.5.1. Catalysis by silver salts. The nature of the catalysis depends upon the experimental conditions, in particular upon the solvent used. In protic solvents the catalysis appears to be mainly heterogeneous, and this situation will be considered first. A well-known example of catalysis by a silver salt is the method used for preparing alcohols of heating an alkyl halide with an aqueous suspension of silver oxide. Other silver salts may be used in place of the oxide; the general features of this type of catalysis can be illustrated by the hydrolysis of ethyl iodide catalysed by silver nitrate [*1.27*]. Under these conditions the reaction between ethyl iodide and silver nitrate is negligible compared with the reaction shown in [*1.27*], and the uncatalysed hydrolysis proceeds at a rate <5 per cent

$$\mathrm{EtI + H_2O + AgNO_3 \longrightarrow EtOH + AgI + HNO_3} \qquad [1.27]$$

of that of the catalysed reaction. By using low concentrations of reactants, and by measuring initial rates of reaction, it was found that the catalysed reaction obeyed the kinetic expression (1.2). Measurements

$$\mathrm{d[HNO_3]/d}t = k\mathrm{[EtI][Ag^+]} \qquad (1.2)$$

of the initial rates were necessary since the reaction was found to be autocatalytic (Austin, Ibrahim and Spiro, 1969).

When finely divided charcoal was added to the reaction mixture a large increase in the rate of hydrolysis was observed. The rate was still proportional to the concentrations of ethyl iodide and silver ion, but the order with respect to the concentration of ethyl iodide was now 0.45 and with respect to the concentration of silver ion 0.3. Such fractional kinetic orders are unusual for homogeneous reactions, but they are consistent with the type of mechanism in which both reactants are adsorbed on the surface of a catalyst, in this case the charcoal. In the present reaction the adsorption itself was shown to be rapid compared with the hydrolysis. The catalytic effects of other solids on reaction [*1.27*] were also investigated; silver halides were found to be most effective, with charcoal and other silver salts being good catalysts, while powdered glass and various metals had little or no effect on the rate of hydrolysis. The effectiveness of a solid surface in promoting the reaction apparently depends partly on its exposed area and partly on its ability to adsorb both reagents well. It seems that the rate-limiting step of this type of reaction involves the heterolysis of the alkyl halide, assisted by a co-ordination between the silver ion and the leaving halide, and that it takes place mainly on the surface of any insoluble silver salt or other solid surface present. The autocatalysis that is observed arises because silver halide is precipitated during the course of the reaction, and this further catalyses the hydrolysis.

A different pattern of behaviour emerges from studies of catalysis in aprotic solvents, under which conditions homogeneous catalysis assumes importance. The function of the silver ion is still to assist the heterolysis of the alkyl halide, although this now occurs readily in solution and does not require the presence of a catalytic surface.

The similarities between the behaviour of the silver-ion catalysed reaction and that of a normal S_N1 reaction suggest that the former type involves a polar transition state. However, it appears unlikely that in aprotic solvents this transition state can be a free carbonium ion. The rate of the reaction between 2-octyl bromide and silver nitrate [1.28] in

$$AgNO_3 + C_6H_{13}-\underset{CH_3}{\overset{Br}{\underset{|}{\overset{|}{C}}}}-H \longrightarrow C_6H_{13}-\underset{CH_3}{\overset{NO_3}{\underset{|}{\overset{|}{C}}}}-H + AgBr + \text{octene} + HNO_3 \qquad [1.28]$$

acetonitrile has been shown to be dependent on the concentration of both the silver ion and the nitrate ion (Pocker and Kevill, 1965). The acceleration produced by the added nitrate ion was too large to be attributed to an S_N2 reaction, and was explained in terms of a mechanism involving anionic assistance in the rate-limiting step [1.29] (cf. the 'push–pull' mechanism). Whilst the nitrate ion is important in the rate-

$$RX + Ag^+ \rightleftharpoons (RX\cdots Ag)^+$$

$$NO_3^- + (RX\cdots Ag)^+ \rightleftharpoons (NO_3^-R^+X^-Ag^+) \qquad [1.29]$$

$$(NO_3^-R^+X^-Ag^+) \longrightarrow \begin{cases} RNO_3 + AgX \\ \text{alkene} + HNO_3 + AgX \end{cases}$$

limiting step, it has virtually no effect on the amount of elimination observed. This suggests that the product-determining step occurs after the rate-limiting step, and in [1.29] it is assumed that the products arise by collapse of the ion quadrupole. An indication that this is a better representation of the reaction intermediate than is a free carbonium ion is that the amount of elimination depends on the leaving halide; the amount of octene formed is 8 per cent for bromide and 3 per cent for chloride.

1.5.2. Catalysis by mercury salts. The use of mercury salts to catalyse nucleophilic substitutions has the advantage over the use of silver salts that in protic solvents the greater solubilities of the former remove the complications due to heterogeneity. However, a new difficulty is introduced, since with mercury salts there may be several effective catalytic species present in solution, e.g. Hg^{2+}, HgX^+, HgX_2, HgX_4^{2-}. For example, in the hydrolysis of alkyl bromides, catalysed by mercuric nitrate, the first-order rate constant is observed to decrease steadily as the reaction proceeds. This may be attributed to the gradual conversion of the ion Hg^{2+} into the less effective catalysts $HgBr^+$ and $HgBr_2$. In spite of the uncertainties introduced into the interpretation of experimental results by this sort of behaviour, it nevertheless appears that, in general, the reactions catalysed by mercury salts exhibit many of the features associated with the normal S_N1 mechanism. The role of the mercury salt is therefore apparently to assist the initial heterolysis of the alkyl halide by co-ordinating with the leaving halide ion.

1.5.3. Catalysis by hydroxylic molecules in aprotic solvents. It was mentioned in an earlier section (§1.2.3) that the reaction between triphenylmethyl chloride and methanol in benzene was of no fixed order in the concentration of methanol, but that it showed an apparent order that rose steadily with increasing concentration of the methanol. One explanation for this is that in non-polar solvents, such as benzene, the polar transition state attracts to itself various numbers of methanol molecules to form a multipolar aggregate. The nucleophilic substitution is then completed within this aggregate. According to this view, one molecule of methanol is involved as the nucleophile, the others merely solvate the polar transition state. Since this solvation can involve an interaction with the departing chloride ion, the role of the methanol may be described as that of an electrophilic catalyst. When the rate of reaction depends on the concentration of methanol to the second order, the reaction might be represented by the scheme [*1.30*]. The dotted lines represent either ionic or covalent interactions.

$$(C_6H_5)_3CCl + CH_3OH \xrightleftharpoons{\text{fast}} (C_6H_5)_3\overset{+}{C}\cdots\overset{-}{Cl}\cdots HOCH_3$$

$$\downarrow \text{slow} \qquad [1.30]$$

$$(C_6H_5)_3COCH_3 \longleftarrow \underset{\underset{H}{|}}{CH_3O}\cdots\overset{+}{C}\cdots\overset{-}{Cl}\cdots HOCH_3$$

An additional example of catalysis is provided by the reactions of t-butyl bromide in nitromethane (Ingold, 1969). When present in low concentrations, the anions Br^-, Cl^-, and NO_2^-, and the hydroxylic molecules water, ethanol, and phenol, all react at the same rate with t-butyl bromide. This, and other observations, suggest that all substitutions occur by the S_N1 mechanism. At higher concentrations of the hydroxylic reagents the rates of substitution are found to rise linearly with reagent concentration. However, the second-order rate constants are not in the order expected on the basis of the nucleophilicities of the hydroxylic reagents, but depend instead on their acidities. This result is easily understood if it is assumed that the hydroxylic reagent is acting as a catalyst and that the transition state for the rate-limiting ionisation resembles (29).

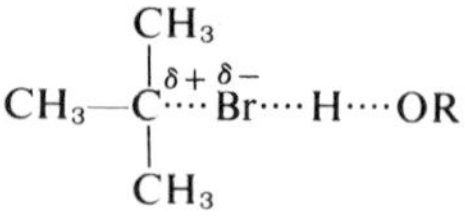

$$CH_3-\overset{\overset{\displaystyle CH_3}{|}}{\underset{\underset{\displaystyle CH_3}{|}}{C}}\overset{\delta+}{\cdots\cdots}\overset{\delta-}{Br}\cdots H\cdots OR$$

(29)

1.5.4. Acid catalysis. In protic solvents acid catalysis is only important in the solvolysis of alkyl fluorides. In aprotic solvents, on the other hand, acid catalysis is observed with alkyl halides in general. A necessary condition, however, is that the solvent be so weakly basic that it does not compete with the alkyl halide for the acid.

The rates of both the radio-chloride exchange and the racemisation of optically active 1-phenylethyl chloride [*1.31*] in nitromethane increase with the concentration of hydrogen chloride (Pocker, Mueller,

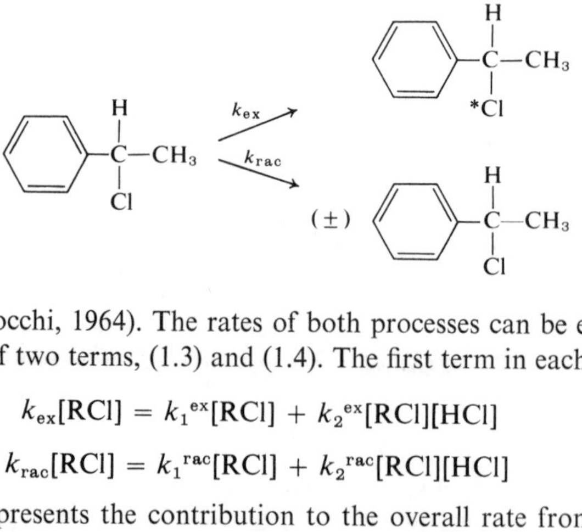

$$C_6H_5-CH(Cl)-CH_3 \xrightarrow{k_{ex}} C_6H_5-CH(^*Cl)-CH_3$$
$$C_6H_5-CH(Cl)-CH_3 \xrightarrow{k_{rac}} (\pm)\ C_6H_5-CH(Cl)-CH_3 \qquad [1.31]$$

Naso and Tocchi, 1964). The rates of both processes can be expressed as the sum of two terms, (1.3) and (1.4). The first term in each of these

$$k_{ex}[RCl] = k_1^{ex}[RCl] + k_2^{ex}[RCl][HCl] \qquad (1.3)$$

$$k_{rac}[RCl] = k_1^{rac}[RCl] + k_2^{rac}[RCl][HCl] \qquad (1.4)$$

equations represents the contribution to the overall rate from an uncatalysed S_N1 process. The second term does not refer to a superposed S_N2 reaction, because it was found that $k_2^{ex} = k_2^{rac}$. For the S_N2 mechanism each act of substitution is accompanied by inversion of configuration and so we should expect to find $2k_2^{ex} = k_2^{rac}$ (the racemisation is complete when half the material is inverted). It was therefore assumed that the molecule of hydrogen chloride present in the rate-limiting step of the second-order component was assisting the heterolysis of the carbon–chlorine bond by hydrogen bonding to the incipient chloride ion, i.e. acid catalysis of an S_N1 process (30).

$$\overset{\delta+}{R}\cdots\overset{\delta-}{Cl}\cdots H\cdots Cl$$

(30)

Molecular hydrogen chloride is known to catalyse other ionic processes in aprotic solvents, for example the Wagner–Meerwein rearrangement of camphene hydrochloride (31) to isobornyl chloride (32) in nitrobenzene [*1.32*].

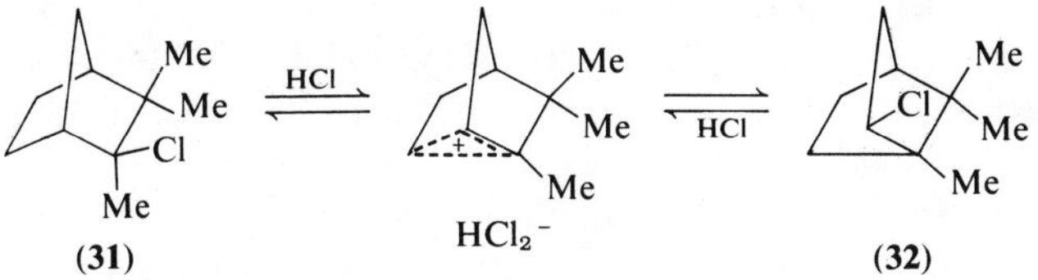

Me Me Cl Me $\underset{}{\overset{HCl}{\rightleftharpoons}}$ Me Me Me HCl_2^- $\underset{HCl}{\rightleftharpoons}$ Cl Me Me Me [*1.32*]

(31) **(32)**

2 Structural and solvent effects

2.1. The structure of the substrate

The reactivity of a substrate towards nucleophilic substitution depends upon its structure. It is useful to be able to predict, at least qualitatively, what effect a change in the structure of the substrate will have on this reactivity. The rate of an S_N1 reaction depends upon the ease with which a substrate forms a carbonium ion, whilst the rate of an S_N2 reaction depends upon the ease with which a nucleophile can effect a direct displacement at a saturated carbon atom. The structural requirements of these two mechanisms are different, so that the effect of a structural change on the reactivity will depend upon the mechanism of the nucleophilic substitution.

For the S_N1 mechanism the effect of the structural change on the rate of formation of the carbonium ion is of interest, and in this connection it is useful to introduce the idea of carbonium ion stability. There are several ways of expressing the relative stabilities of carbonium ions, one of which is to use measurements of the rates of ionisation of the parent compounds (Bethell and Gold, 1967). This basis for comparison rests on the assumption that for a given type of parent compound a more stable ion will be formed more rapidly. This means that we may rationalise the effects of structural changes on the rates of S_N1 reactions in terms of how these changes can be expected to influence the stabilities of the corresponding carbonium ions.

A comparison of rates actually gives the difference between the Gibbs free energies of activation, but it provides a measure of carbonium ion stability if the following assumptions are made: that the difference between the Gibbs free energies of activation is due largely to the difference between the free energies of the transition states, and that the free energies of the transition states parallel those of the corresponding carbonium ions (fig. 2.1).

The stability of a carbonium ion is often referred to when discussing the reactivity of that species towards a nucleophile. However, the rate at which a carbonium ion reacts with a nucleophile is determined by the difference in free energy between the carbonium ion and the transition state for the combination; the more reactive the carbonium ion, the smaller this free energy difference. It should be noted that the reactivity is not therefore related to the stability of the carbonium ion, as this latter property was defined above. It remains a fact, nevertheless, that in

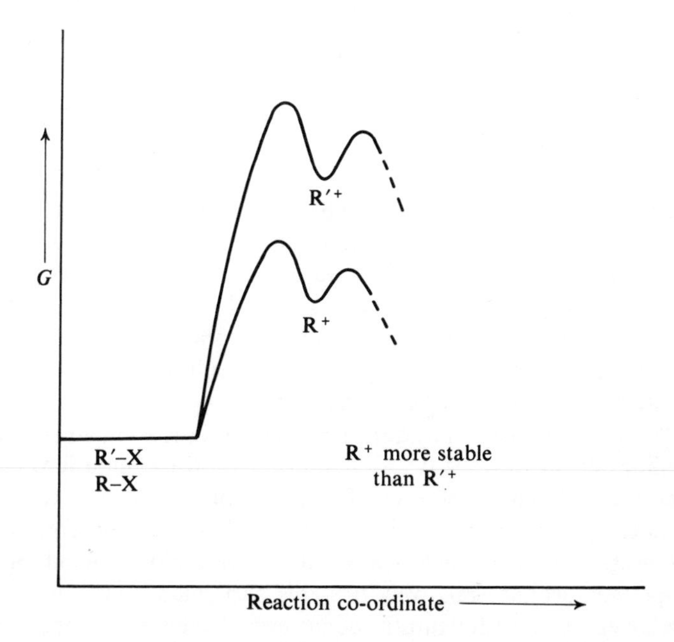

Fig. 2.1 Free energy diagram representing the formation of carbonium ions of different stabilities.

general the less stable carbonium ions are the more reactive, but it should be emphasised that these two properties need not necessarily be so related.

The effect of a structural change on the rate of an S_N2 reaction is more difficult to predict, because in this case both bond making and bond breaking are important in the transition state, and these processes have opposite electronic requirements. It is possible, though, to make a general comment about the steric requirements of this type of mechanism. Increased branching in the structure should tend to decrease the rate of substitution, since the nucleophile should experience greater difficulty in attacking the reacting carbon atom as this becomes shielded by bulky substituents.

2.1.1. Alkyl groups. There are three types of structure which are usually referred to in a discussion of the mechanisms of nucleophilic substitutions; primary (33), secondary (34), and tertiary (35), which

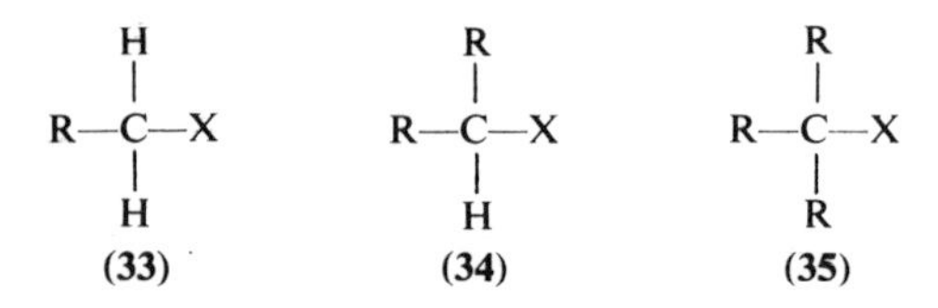

differ from one another in the number of alkyl groups that are attached directly to the reacting carbon atom (the α-carbon atom). As a general rule, primary alkyl compounds tend to undergo nucleophilic substitutions by the S_N2 mechanism and tertiary alkyl compounds by the S_N1 mechanism. The secondary alkyl compounds show a more variable type of behaviour and are therefore more difficult to classify; they are usually placed in the borderline category.

This tendency to react by different mechanisms is a consequence of the stabilities of the various types of simple alkyl carbonium ions derived from the different structures. Tertiary compounds give reasonably stable carbonium ions and these compounds readily undergo nucleophilic substitution by the S_N1 mechanism. On the other hand, primary compounds produce relatively unstable carbonium ions, and in such cases an easier route for substitution is provided by the S_N2 mechanism, in which the heterolysis of the bond to the leaving group is assisted by the interaction between the reacting carbon atom and the incoming nucleophile. With secondary compounds there is a finer balance between unassisted ionisation and direct displacement.

The reaction conditions can be adjusted so that all three types of compounds react by the S_N2 mechanism, but the corresponding situation in which all react by the S_N1 mechanism has probably not yet been achieved (§4.3.5). The solvolytic displacements on tertiary compounds usually take place by the S_N1 mechanism and this structural type provides a convenient one with which to illustrate some of the effects of structural changes.

With tertiary alkyl compounds, provided that the structures do not become too ramified, increased alkyl substitution has only a small effect on the rate of reaction (table 2.1). This result appears to be consistent with the stabilisation of the transition state by the inductive effect of the alkyl groups (Streitwieser, 1956). When the groups R_1, R_2 and R_3 become excessively bulky, the rates of solvolysis become too large to be attributed to a stabilisation of the transition state by inductive effects alone, and in such cases steric factors are assumed to be important.

TABLE 2.1 *The rates of solvolysis of tertiary alkyl chlorides,* $R_1R_2R_3CCl$, *in 80 per cent ethanol at 25 °C*[a]

R_1	R_2	R_3	Relative rate
CH_3	CH_3	CH_3	1
C_2H_5	CH_3	CH_3	1.65
C_2H_5	C_2H_5	CH_3	2.58
C_2H_5	C_2H_5	C_2H_5	3.02
n-C_3H_7	CH_3	CH_3	1.58
n-C_4H_9	CH_3	CH_3	1.43
t-C_4H_9	CH_3	CH_3	1.21

[a] Data from Streitwieser (1956).

These arise from unfavourable non-bonding interactions (§2.1.3) among the bulky groups attached to the α-carbon atoms in the starting material, which may be partly relieved by the change in geometry that accompanies ionisation [*2.1*]. This effect produces an enhanced rate of

$$\mathrm{C{-}X} \longrightarrow \overset{+}{\mathrm{C}} + \mathrm{X^-} \qquad [2.1]$$

reaction through steric acceleration of ionisation. The non-bonding interactions may also be eased by molecular rearrangement and by elimination, and both processes frequently occur concurrently with substitution.

With highly ramified structures elimination may occur to the exclusion of substitution, but the rates of reaction may still give information about the ionisation process. The products obtained from a series of t-alkyl *p*-nitrobenzoates (table 2.2) were shown to be olefins, formed with varying amounts of rearrangement, but the last two compounds mentioned were found to give products with predominantly unrearranged carbon skeletons (Bartlett and Tidwell, 1968). These results were taken as showing that the rates were subject to steric acceleration of ionisation and that no important contribution to the enhanced rates was caused by the participation of a rearranging alkyl group (§5.2.2).

TABLE 2.2 *The rates of solvolysis of t-alkyl p-nitrobenzoates,* $R_1R_2R_3COpNB$, *in 60 per cent dioxane at 40 °C*[a]

R_1	R_2	R_3	Relative rate
Me	Me	Me	1
t-Bu	t-Bu	t-Bu	13 500
neopentyl	t-Bu	t-Bu	19 400
neopentyl	neopentyl	t-Bu	68 000
neopentyl	neopentyl	neopentyl	560

[a] Data from Bartlett and Tidwell (1968).

The introduction of the third neopentyl group, to give trineopentylcarbinyl *p*-nitrobenzoate, is accompanied by a large decrease in rate. Apparently three neopentyl groups about a central carbon atom are subject to less strain than any combination of neopentyl and t-butyl groups, a conclusion supported by the construction of molecular models.

In order to determine the effect on the rate of an S_N1 reaction of increasing the number of alkyl groups attached directly to the α-carbon atom, it is necessary to compare the reactions of tertiary and secondary (or secondary and primary) compounds. Such a comparison will only give a meaningful estimate of the substituent effect if both compounds react by the same mechanism. As mentioned above, it is not certain whether, under the usual reaction conditions, primary and secondary compounds react by the S_N1 mechanism.

A secondary compound which apparently undergoes solvolysis by the S_N1 mechanism is 2-adamantyl bromide (36*a*). The rates of solvolysis of (36*a*) and the corresponding tertiary compound (36*b*) have been

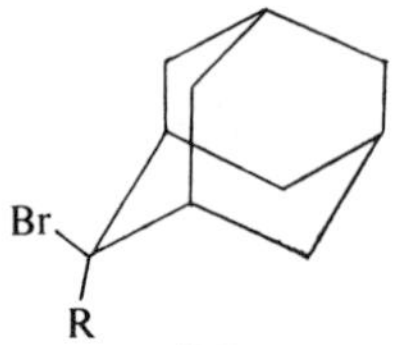

Br R

a R = H
b R = Me

(36)

measured in 80 per cent ethanol and in acetic acid (Fry, Harris, Bingham and Schleyer, 1970) and in both solvents (36*b*) is more reactive than (36*a*) by a factor of 10^8. The suggestion has been made that this factor represents the true substituent effect of an α-methyl group for the S_N1 mechanism, and that smaller rate ratios indicate that like processes are not being compared. By this criterion very few of the simple primary or secondary alkyl compounds react by a pure S_N1 mechanism. The large increase in the rate of solvolysis caused by the substitution of a methyl group for a hydrogen atom at the α-carbon atom is too large to be due to an inductive effect alone, and is unlikely to be the result of a steric effect. It seems best attributed to the operation of hyperconjugation [*2.2*], which stabilises the carbonium ion by effectively dispersing the positive charge.

$$R_2C^+\text{—}CH_3 \longleftrightarrow R_2C\text{=}CH_2\ H^+ \longleftrightarrow \text{etc.} \qquad [2.2]$$

The effect of α-methyl substitution on the rates of solvolysis of some simple alkyl compounds is shown in table 2.3. A large increase in the

TABLE 2.3 *The rates of solvolysis of alkyl bromides,* $R_1R_2R_3CBr$, *under various reaction conditions*[a]

R_1	R_2	R_3	Relative rates			
			EtOH, 55 °C	50% EtOH, 55 °C	H_2O, 50 °C	HCOOH, 100 °C
H	H	H	1	1	1	1
CH_3	H	H	0.39	0.58	0.95	1.7
CH_3	CH_3	H	0.28	1.66	11	45
CH_3	CH_3	CH_3	320	~2.8×10^4	1.14×10^6	~1.7×10^8

[a] Data from Streitwieser (1962).

rate is only observed with t-butyl bromide, the other compounds showing relatively small differences in reactivity. This behaviour is almost certainly due to the fact that the different compounds do not react by the same mechanism, a conclusion supported by the way in which the relative rates change with the reaction conditions. Of the solvents reported, ethanol should be the least favourable and formic acid the most favourable for the observation of an S_N1 mechanism.

The effect of structural changes in the alkyl group on the rates of S_N2 reactions are quite different from those noted above for S_N1 reactions. In general, an increase in branching of the alkyl structure leads to a decrease in the rate of reaction (table 2.4). When the substitution of methyl for hydrogen occurs at the α-carbon atom there is a

TABLE 2.4 *The effect of the alkyl group on the rate of the displacement* $RBr + Cl^- \rightarrow RCl + Br^-$ *in dimethylformamide at 25 °C*[a]

R	k_2 (l mol^{-1} s^{-1})
Me	5.00×10^{-1}
Et	1.35×10^{-2}
i-Pr	2.45×10^{-4}
t-Bu	1.10×10^{-5}
n-Pr	9.32×10^{-3}
i-Bu	4.46×10^{-4}
neopentyl	8.50×10^{-8}

[a] Data from Cook and Parker (1968).

steady decrease in reaction rate (cf. the series, Me, Et, i-Pr, t-Bu). Similar substitution at the β-carbon atom, however, gives a different pattern of behaviour; successive methyl substitution is progressively more effective at reducing the reaction rate, the large decrease occurring with neopentyl being particularly striking. The operation of steric factors is believed to be partly responsible for this type of behaviour.

The approach of the nucleophile, N, to the transition state distance (37) leads to an increase in the non-bonding interactions (§2.1.3),

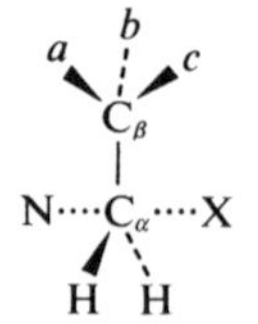

a, b, c — C_β — N····C_α····X (H, H)

(37)

which become more severe as the hydrogen atoms at positions *a*, *b* and *c* are successively replaced by methyl groups. The introduction of one methyl group has little effect on the rate of reaction, since, by suitable orientation of the $C_\beta abc$ moiety, serious non-bonding interactions may be avoided. The same is true, although to a lesser extent, when two methyl groups are present, but the introduction of the third methyl group leads necessarily to unavoidable non-bonding interactions.

The contributions made by non-bonding interactions to the energies of activation have been calculated for the series of structures listed in table 2.4 (Ingold, 1969). On comparing the calculated values with the observed energies of activation it was concluded that both electronic and steric effects are important in determining the reactivities, but that the latter effect becomes dominant in the more ramified structures.

2.1.2. Aryl groups. An aryl group attached to the α-carbon atom assists a reaction occurring by the S_N1 mechanism by allowing the developing positive charge to be distributed over the conjugated system [*2.3*]. The greater stability of aryl relative to alkyl carbonium ions is

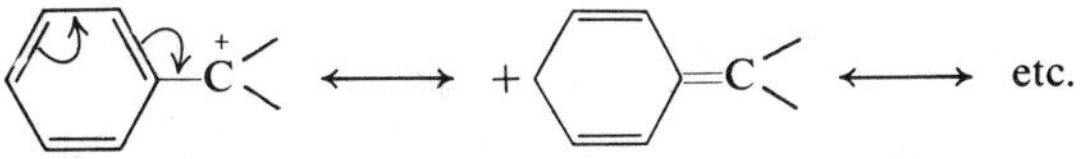

etc. [*2.3*]

indicated by the fact that limiting S_N1 behaviour is frequently observed with secondary aryl systems, e.g. (38) and (39), and in some cases with primary systems (40). In the preceding section it was seen that limiting S_N1 reactions are usually observed only with tertiary alkyl systems.

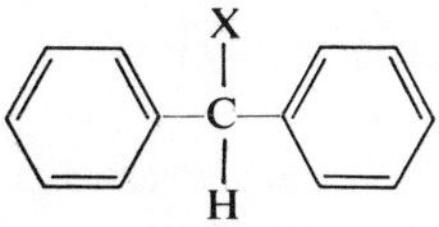

(38) 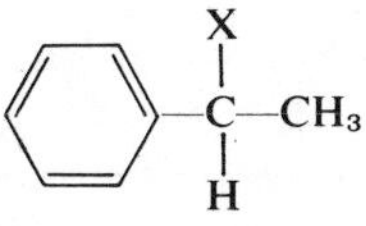 **(39)** 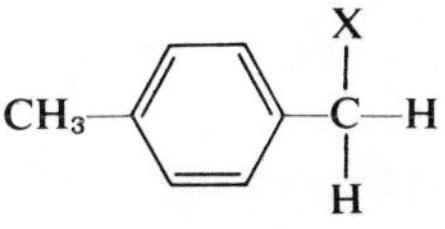 **(40)**

Compounds with α-phenyl substituents have been widely studied, particularly when additional substituents have been present at the *meta*- or *para*-positions of the phenyl ring. As expected, electron donating groups further enhance the rates of S_N1 reactions and, in general, the effects of ring substituents are quantitatively correlated by the Hammett $\sigma\rho$ relation, when σ^+ values are used for the substituent constants (§3.6.1). There appears to be no reliable estimate of the activating effect of an α-phenyl substituent, but it is probably of the same order of magnitude as that of an α-methyl group (see p. 31).

In addition to the effects of ring substituents, the S_N1 reactivity of a system may be increased by annellation, which allows further delocalisation of the positive charge over the additional rings. These systems are amenable to MO calculations and provide a convenient series with which to test the various theoretical approaches. The aim of any calculation is to estimate the change in energy on going from the reactant molecule to the transition state, but this is usually approximated to the energy change, ΔE, for process [*2.4*], in which the hydrocarbon is used as a model for the reactant $ArCH_2X$, and the corresponding carbonium ion is used to represent the transition state. If it is assumed that for the

$$ArCH_3 \longrightarrow ArCH_2^+ \qquad [2.4]$$

series of compounds studied the entropy of activation remains constant and the enthalpies of solvation of the carbonium ions are similar, then the rate of the S_N1 reaction is given by (2.1), in which A is a constant.

$$-2.303\ RT \log k = A + \Delta E \qquad (2.1)$$

If one system is chosen as a reference, then relative reactivities should be given by (2.2). Various methods have been used to calculate values of

$$2.303\ RT \log k_{rel} = -\Delta\Delta E \qquad (2.2)$$

ΔE, and thence of $\Delta\Delta E$, and the results indicate that the latter values usually show a satisfactory correlation with observed relative reactivities (Streitwieser, Hammond, Jagow, Williams, Jesaitis, Chang and Wolf, 1970). A typical correlation is shown in fig. 2.2, in which the relative rates of acetolysis, measured at 40 °C, for a series of arylmethyl tosylates† (benzyl is the reference group) are plotted against the corresponding values of $-\Delta\Delta E$.

Aryl substitution often has little effect on the rates of S_N2 reactions; thus methyl chloride and benzyl chloride both undergo the Finkelstein reaction with iodide ion, [*2.5*] and [*2.6*], in acetone at about the same rate. The α-phenyl group might be expected to introduce at least some

$$CH_3Cl + I^- \longrightarrow CH_3I + Cl^- \qquad [2.5]$$

$$C_6H_5\text{—}CH_2Cl + I^- \longrightarrow C_6H_5\text{—}CH_2I + Cl^- \qquad [2.6]$$

unfavourable non-bonding interactions into the S_N2 transition state of [*2.6*], which should lead to a decrease in rate, hence the above result indicates that some other factor is operating. It seems reasonable to assume that this additional factor is electronic in origin.

The rate of an S_N2 reaction may be increased by either an electron-donating or an electron-withdrawing substituent at the *para*-position of

† *p*-Toluenesulphonate, —OTs.

an α-phenyl group, but such effects are usually quite small (Streitwieser, 1962). This behaviour can be understood if it is assumed that the relative importance of bond making and bond breaking at the transition

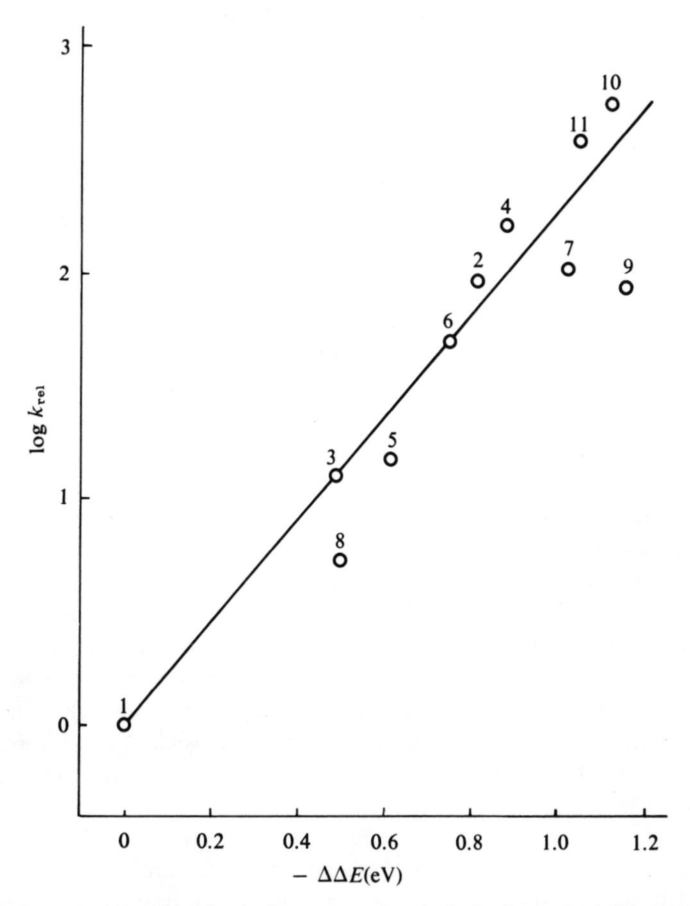

Fig. 2.2 Correlation of relative rates of acetolysis for polycyclic tosylates, $ArCH_2OTs$, with values of $\Delta\Delta E$. 1, phenyl; 2, 1-naphthyl; 3, 2-naphthyl; 4-2-anthracyl; 5, 2-phenanthryl; 6, 3-phenanthryl; 7, 9-phenanthryl; 8, 2-pyrenyl; 9, 4-pyrenyl; 10, 3-fluoranthryl; 11, 8-fluoranthryl. Data from Streitwieser *et al.* (1970).

state varies with the experimental conditions. The S_N2 transition state, which is normally represented by (41) may, in fact, be closer to one of the extremes represented by the tight structure (42) or the loose structure

(43). In (42) the C–N bond is formed almost completely before any appreciable breaking of the C–X bond occurs, with a corresponding increase in electron density at the reacting carbon atom. A transition state of this structure should be stabilised by electron-withdrawing substituents. At the other extreme (43), C–X bond breaking runs ahead

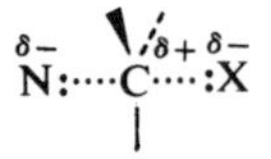

$\overset{\delta-}{N}{:}\cdots\overset{\delta+}{C}\cdots{:}\overset{\delta-}{X}$ (41)

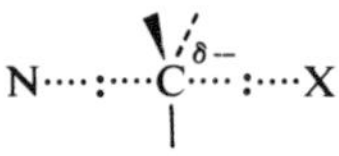

$N\cdots{:}\cdots\overset{\delta-}{C}\cdots{:}\cdots X$ (42)

$N{:}^{-}\quad C^{+}\quad {:}X^{-}$ (43)

of bond formation so that the reacting carbon atom now has appreciable carbonium ion character. A transition state of this structure should be stabilised by electron-donating substituents.

The nature of the transition state is determined in part by the structure of the substrate and in part by the properties of the nucleophile and the leaving group. It is therefore possible that a given ring substituent might be activating in some reactions and deactivating in others, although the mechanism remains S_N2 throughout.

2.1.3. Steric effects and molecular strain. We have seen in the preceding discussion that interactions among atoms or groups not connected by chemical bonds can influence the rates of both S_N1 and S_N2 reactions. In the former case the relief of severe non-bonding interactions in the substrate can provide the driving force for ionisation, while in the latter case the increased non-bonding interactions in the transition state can seriously impede the approach of the nucleophile. The forces between a pair of non-bonded atoms remain weak until a certain critical separation (the sum of the van der Waals radii) is reached, within which they become strongly repulsive. Non-bonding energy is present in all molecular structure, but it usually has no significant effect on reaction rates. When, however, structural features of either the reactant or the transition state force non-bonded atoms to separations which are less than the sums of their van der Waals radii, then kinetic effects are observed which are usually attributed to the operation of steric factors.

Non-bonding interactions are a source of strain which contributes to the total energy of a molecule. The existence of strain and its effect on molecular structure is an important factor influencing chemical reactivity. The most favourable conformation adopted by a molecule is that arrangement which has the lowest total energy, i.e. which involves the least strain; the total energy of a molecule is conveniently expressed as a sum, (2.3), giving the contributions from the various sources of strain.

$$E_{\text{total}} = E_{\text{stretch}} + E_{\text{bend}} + E_{\text{torsion}} + E_{\text{non-bonding}} \tag{2.3}$$

The first two terms, E_{stretch} and E_{bend}, represent the energy arising from the stretching and compressing of bonds, and the distorting of bond angles. As mentioned above, $E_{\text{non-bonding}}$ arises from the repulsive interactions of non-bonded atoms. The term E_{torsion} is more difficult to define, but it is best thought of as the energy arising from the resistance to the rotation of one part of the structure relative to another part (it is calculated from the height of the barrier to rotation).

The total energy calculated by (2.3) is sometimes called the strain energy, but it is more usual to define strain as the difference in energy between the molecule of interest and some model compound which is assumed to be 'strain-free'. Some authors also refer to the total energy as the steric energy, but it would seem preferable to restrict the use of the term steric to those effects which depend directly upon non-bonding interactions.

The ability to be able to calculate the structures and energies of molecules provides an additional probe with which to investigate the effects of structure on the reactivities of organic molecules. The approach is based on that suggested by Westheimer (1956), in which the energies and the preferred conformations are found by adjusting the geometry of the molecule until a minimum energy is obtained. This method of conformational analysis has proved feasible only since the application of digital computers to the calculations. Standard methods are now available which enable the terms in (2.3) to be calculated with a moderate degree of accuracy (Allinger, Tribble, Miller and Wertz, 1971).

For this method to be of use in estimating the reactivity of a molecule, it is necessary to calculate the difference in energy between the ground state and the transition state. This requires a conformational analysis to be performed on a suitable model for the transition state, a procedure which introduces some uncertainty into the calculation.

2.1.4. Cycloalkyl systems. The reactivity of a cycloalkyl compound shows a marked dependence upon the size of the ring (Brown and Ichikawa, 1957). This correlation is related to the presence of strain, and to the way in which the strain varies with the structure. The energies of the cycloalkanes (C_4–C_{12}) have been calculated by conformational analysis, and the estimates of the strain energies (relative to that of cyclohexane) so derived are in good agreement with those obtained from experimental heats of combustion (Allinger, Tribble, Miller and Wertz, 1971). The conformational analysis also shows the way in which the contributions of the individual terms of (2.3) vary with ring size (fig. 2.3).

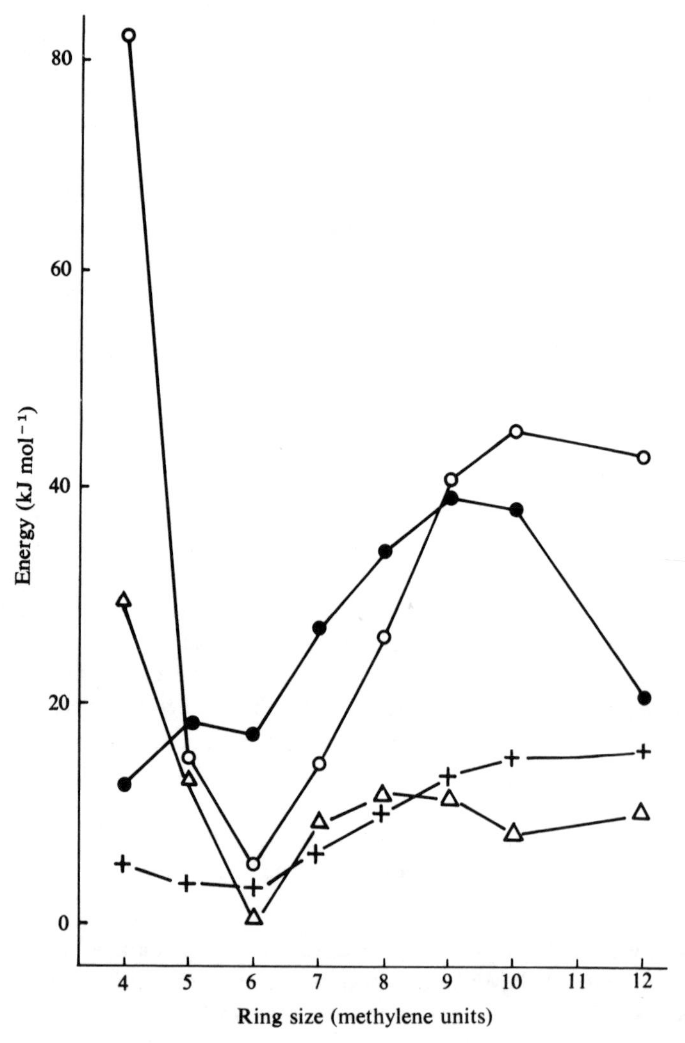

Fig. 2.3 Contributions to the total energy, E_{total}, in the C_4–C_{12} rings. + $E_{stretch}$, ○ E_{bend}, △ $E_{torsion}$, ● $E_{non-bonding}$. Data from Allinger *et al.* (1971).

Although strain in the substrate has important consequences, of greater interest from the point of view of reactivity is the change in the strain energy which accompanies formation of the transition state. These changes have been calculated for several cycloalkyl arenesulphonates using simple semiempirical methods to estimate the contributions from changes in angle strain, torsional strain and non-bonding interactions, and assuming S_N1 behaviour (Schleyer, 1964). The relative reaction rates so calculated agree very well with the observed rates of acetolysis at 25 °C (relative to that of cyclohexyl) for the C_5–C_{15} compounds (fig. 2.4). The disagreement between the observed and

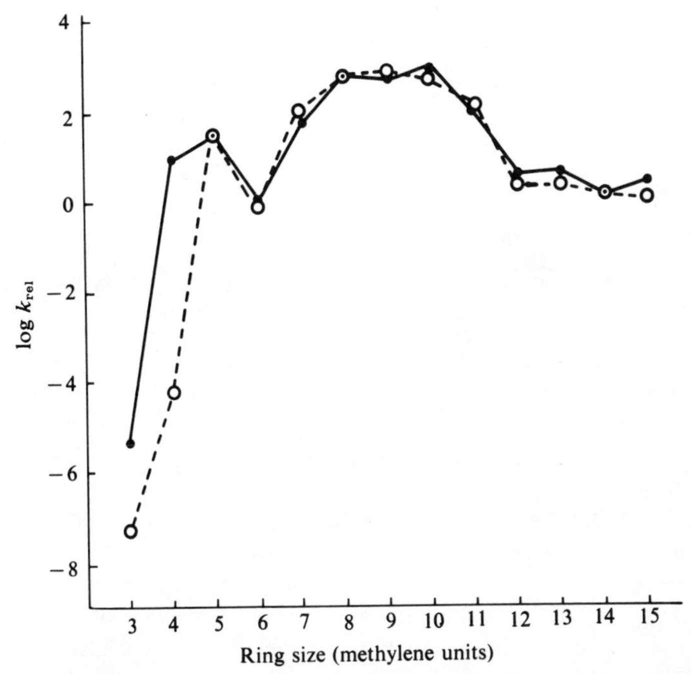

Fig. 2.4 The effect of ring size on the rates of acetolysis at 25 °C of cycloalkyl arenesulphonates. ● observed, ○ calculated. Data from Schleyer (1964).

calculated rates for both cyclopropyl and cyclobutyl is most probably connected with the fact that with both compounds rearrangement accompanies acetolysis [*2.7*] and [*2.8*]. The initial ionisation in both cases is thought to lead directly to a bridged carbonium ion which is more stable than the corresponding cyclopropyl or cyclobutyl cation.

$$\text{cyclopropyl-OTs} \longrightarrow [CH_2{\cdots}CH{\cdots}CH_2]^+ \longrightarrow CH_2{=}CH.CH_2OAc \quad [2.7]$$

$$\text{cyclobutyl-OTs} \longrightarrow H_2C{\cdots}CH(CH_2){\cdots}CH_2^+ \longrightarrow \text{cyclopropyl-}CH_2OAc \text{; cyclobutyl-}OAc \text{; } CH_2{=}CH.CH_2.CH_2.OAc \quad [2.8]$$

The effect of ring size on the rates of S_N2 reactions is indicated by the results for the halide exchange reactions of the C_3–C_8 cycloalkyl bromides (table 2.5). The general pattern of behaviour is similar to that

TABLE 2.5 *Relative rates of reaction for the halide exchange* $RBr + I^- \rightarrow RI + Br^-$ *in acetone at 25 °C*[a]

R	k_{rel}
Cyclopropyl	–[b]
Cyclobutyl	0.75
Cyclopentyl	160
Cyclohexyl	1.0
Cycloheptyl	98
Cyclo-octyl	22

[a] Data from Streitwieser (1962).
[b] No measurable reaction after 8 h with 0.01M KI at 200 °C.

observed with the acetolysis results (fig. 2.4), and indicates that the same factors are important in determining the reactivities of the cycloalkyl systems when these react by either the S_N1 or the S_N2 mechanisms. In reactions occurring by the latter mechanism there will, of course, be additional non-bonding interactions to consider in the transition state arising from the presence of the nucleophile.

2.1.5. Bridgehead systems. Conformational analysis calculations have provided some insight into the factors controlling the reactivities of bridgehead positions. Compounds such as apocamphyl chloride (44) are extremely reluctant to undergo nucleophilic substitution reactions. The bridged structure prevents substitution occurring by the S_N2 mechanism and inhibits reaction by the S_N1 mechanism. This latter circumstance is

Cl

(44)

related to the difficulty of forming a carbonium ion at the bridgehead position.

The conformation of minimum energy for a carbonium ion is thought to have a planar geometry about an sp^2 hybridised carbon atom (Fort and Schleyer, 1966). Any change from this preferred geometry will lead to an increase in energy, and therefore to a loss of stability, of the carbonium ion. The ease of formation of a carbonium ion at a bridgehead position, and therefore the ease of substitution by the S_N1 mechanism of substitution, depends upon the ability of the cyclic structure to accommodate a planar, or nearly planar, geometry at the bridgehead. This depends upon the size of the rings, and in general the greater rates of nucleophilic substitution are associated with the larger ring systems (table 2.6). It should be noted that not all bridgehead compounds are so unreactive towards nucleophilic substitution as apocamphyl chloride (or the 1-norbornyl system); 3-homoadamantyl bromide, for example, is about as reactive as t-butyl bromide.

The energies and preferred conformations of bridgehead compounds, and those of the corresponding carbonium ions, may in principle be calculated by conformational analysis. This would allow the increase in total energy, ΔE, involved in the formation of a carbonium ion at a bridgehead to be calculated. Calculations of this type have been performed for several of the compounds included in table 2.6 (Gleicher and Schleyer, 1967), and the observed reactivities appear to be satisfactorily accounted for in terms of the variation in the calculated values of ΔE (fig. 2.5).

In addition to giving the increase in the total energy, the calculations show how the individual terms of (2.3) change on going from reactant to carbonium ion. Such studies indicate that it is the increase in the bond angle term, E_{bend}, which is largely responsible for the increases in total energy. The flattening of the bridgehead, in an attempt to produce a planar carbonium ion, leads to distortions in the bond angles of the whole molecule and to a corresponding increase in energy. Bond angle strain appears to be particularly acute in the 1-norbornyl cation and, indeed, this system is remarkably unreactive (table 2.6, cf. apocamphyl chloride). The terms $E_{torsion}$ and $E_{non\text{-}bonding}$ also increase on formation of the carbonium ion and in some cases they too make a significant contribution to ΔE.

TABLE 2.6 *Relative rates of solvolysis of bridgehead bromides in 80 per cent ethanol at 25 °C*[a]

Compound		k_{rel}
Me_3CBr	t-Butyl	1.0
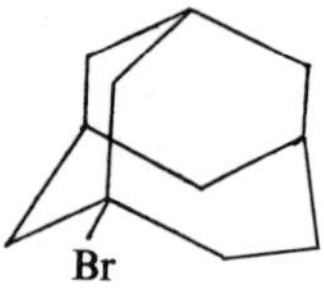 Br	3-Homoadamantyl	0.5
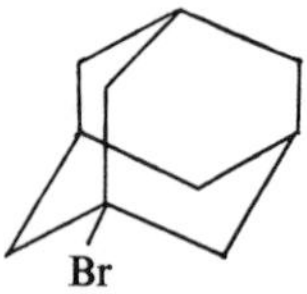 Br	1-Adamantyl	10^{-3}
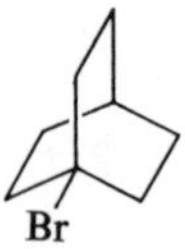 Br	1-Bicyclo[2,2,2]octyl	10^{-6}
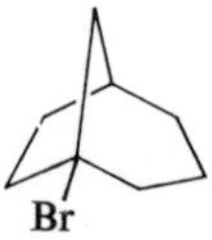 Br	1-Bicyclo[3,2,1]octyl	10^{-6}
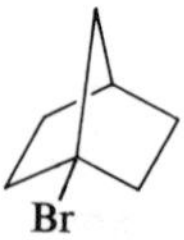 Br	1-Norbornyl	10^{-13}
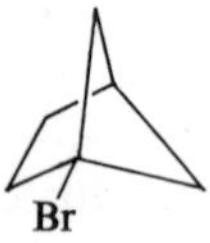 Br	1-Bicyclo[2,1,1]hexyl	10^{-6}

[a] Data from Fort and Schleyer (1966).

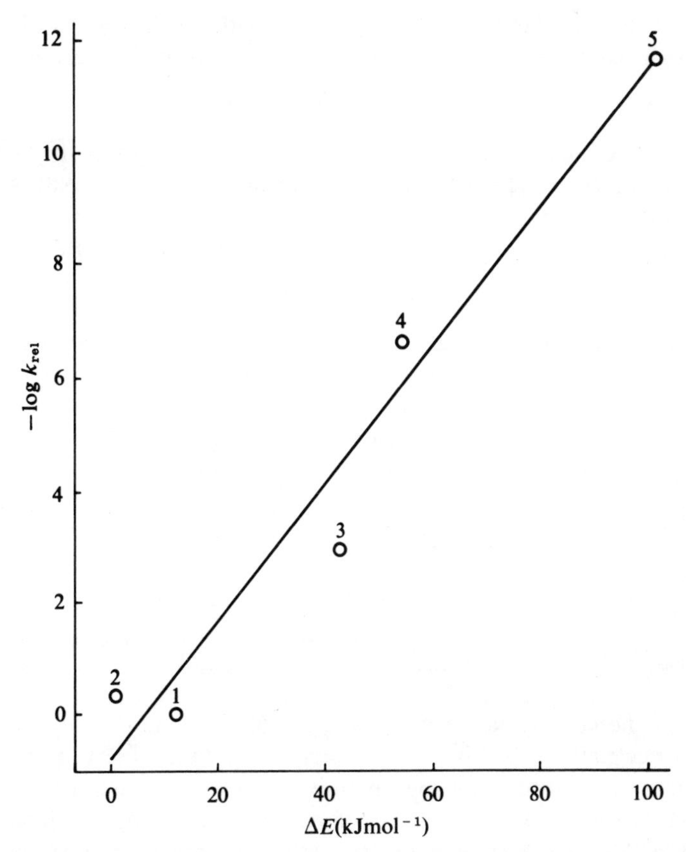

Fig. 2.5 Correlation between calculated values of ΔE and observed relative rates of solvolysis for several bridgehead compounds. 1, t-butyl; 2, 3-homoadamantyl; 3, 1-adamantyl; 4, 1-bicyclo[2,2,2]octyl; 5, 1-norbornyl. Data from Gleicher and Schleyer (1967).

2.2. Nucleophilic reactivity

Before discussing the factors which influence the reactivity of a nucleophile it is necessary to make a distinction between the terms nucleophilicity and basicity. Both refer to the tendency of a reagent to supply an unshared pair of electrons to form a covalent bond. But whereas basicity is used in a thermodynamic sense and applies to equilibrium conditions, nucleophilicity refers to kinetic phenomena and is therefore associated with reaction rates.

As commonly used, the term basicity is simply a measure of the thermodynamic affinity of a compound for the hydrogen ion [*2.9*], although basicities towards elements other than the hydrogen ion may

$$N: + H_3O^+ \rightleftharpoons NH^+ + H_2O \qquad [2.9]$$

also be defined. The inherent reactivity of a nucleophile, its nucleophilicity, may be defined as the kinetic affinity of the nucleophile for a saturated carbon atom; a measure of this is given by the rate constant for a displacement reaction of the type [*2.10*] (see p. 49). Reaction centres other than

$$N: + RX \longrightarrow RN + X: \qquad [2.10]$$

a saturated carbon atom may also be used to define nucleophilicity, but these will not be considered in the present discussion.

2.2.1. Basicity. The rates of the S_N2 reactions of phenacyl bromide with *para*-substituted phenoxide ions [*2.11*] are linearly related to the

$$C_6H_5\text{—}COCH_2Br + {}^-O\text{—}C_6H_4\text{—}Y \longrightarrow C_6H_5\text{—}COCH_2O\text{—}C_6H_4\text{—}Y + Br^- \qquad [2.11]$$

pK_as of the corresponding phenols (fig. 2.6). Similar correlations between nucleophilic reactivity and basicity have been observed for the reactions of substituted anilines, pyridines, phosphines, thiophenoxides and carboxylates. Within each of the series mentioned, the different nucleophiles have very similar structures, the variations in structure being at positions well removed from the reaction centre.

When the structure of the nucleophile is not restricted to such a limited range of variation, a linear correlation between nucleophilic reactivity and basicity is no longer observed. It still remains true, however, that the strongest bases tend to be the most effective nucleophiles, but only then for those cases in which the reacting atom remains the same, as in the series of oxygen nucleophiles, $OH^- > OPh^- > OAc^- > H_2O > ClO_4^-$, whose nucleophilicities and basicities decrease in the order indicated. Any larger structural variation leads to a complete loss of correlation. The thiophenoxide anion is a much more effective nucleophile than the phenoxide anion, although it is the weaker base of the two. Such observations indicate that factors other than basicity are important in determining nucleophilic reactivity.

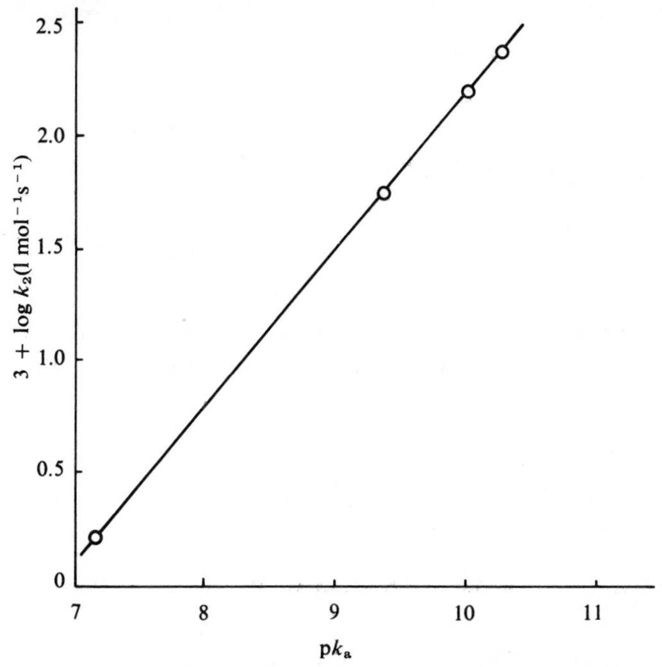

Fig. 2.6 Correlation between nucleophilic reactivity and basicity for phenoxide ions. Rates for the S_N2 reactions with phenacyl bromide at 30 °C in 50 per cent aqueous acetone. Data from Okamoto, Kushiro, Nitta and Shingu (1967).

2.2.2. Polarisability. Nucleophilic reactivity appears to increase with increasing atomic number for elements in the same group of the periodic table. This applies both to simple anions, e.g. $I^- > Br^- > Cl^- > F^-$, and to more complex reagents, e.g.

$$C_6H_5Se^- > C_6H_5S^- > C_6H_5O^-$$

(cf. table 2.7). The greater reactivity of a larger anion is assumed to be a consequence of the greater ease with which the valence electrons are distorted (polarised) by the electrophilic centre of the substrate being attacked. Two effects may be distinguished (Edwards and Pearson, 1962). The first is the polarisation of the bonding electrons from the nucleophile towards the substrate, which allows better electrostatic interactions and promotes bond formation. The second is the polarisation of the non-bonding electrons on the nucleophile away from the substrate which tends to reduce electrostatic repulsions between the incoming nucleophile and the leaving group.

The observed order of nucleophilic reactivity for many reactants can be rationalised in terms of basicity and polarisability factors. For reactions in protic solvents, for example, the following order of reactivity is commonly observed:

$$C_4H_9S^- > C_6H_5S^- > S_2O_3^{2-} > (NH_2)_2CS > I^- > CN^- > SCN^- > OH^- > N_3^- > Br^- > C_6H_5O^- > Cl^- > C_5H_5N > CH_3CO_2^- > H_2O$$

When the reacting atom is the same, the relative reactivities of the nucleophiles are determined by their basicities; otherwise, the polarisability factor appears to be the more important in determining the order of reactivity.

2.2.3. Steric effects. Quinuclidine (45) is more reactive than triethylamine (46) towards methyl iodide in nitromethane at 25 °C, although the basicities of both amines are nearly the same (Brown and Eldred, 1949). The difference in reactivity becomes progressively larger as the substrate is changed, first to ethyl iodide, then to isopropyl iodide. These

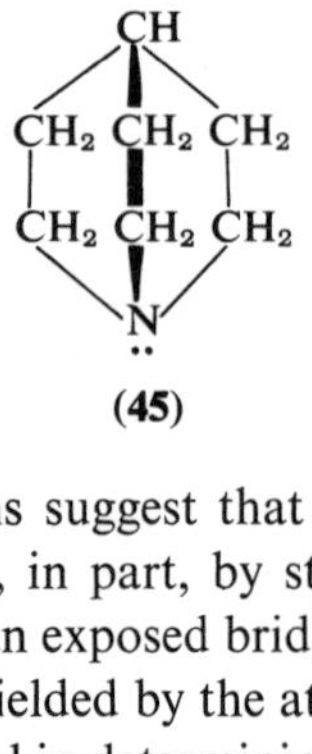

(45) 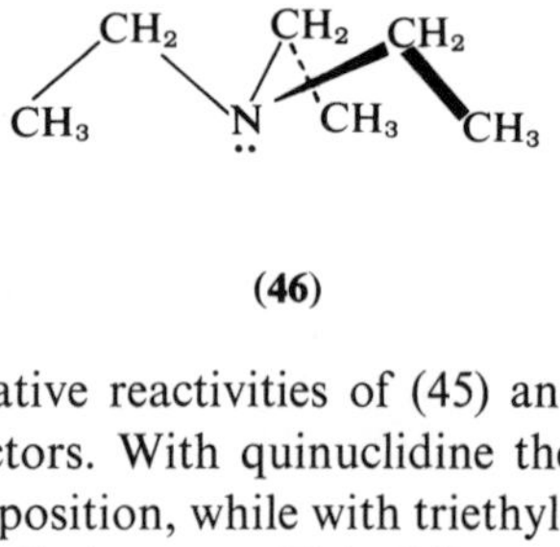 (46)

observations suggest that the relative reactivities of (45) and (46) are determined, in part, by steric factors. With quinuclidine the nitrogen atom is at an exposed bridgehead position, while with triethylamine it is partially shielded by the attached alkyl groups. This difference becomes more critical in determining the relative reactivity as the structure of the alkyl halide becomes more ramified.

2.2.4. The alpha effect. In the nucleophilic displacement of bromide ion from benzyl bromide in 50 per cent aqueous acetone at 25 °C [*2.12*], the hydroperoxide anion is about thirty-five times more reactive than the hydroxide ion, even though the latter is the stronger base by a factor of 10^4. This difference in rate cannot easily be explained in terms of the polarisabilities, since there is no reason to suppose that they should be very different for the two nucleophiles. Other examples of nucleophiles which show enhanced reactivities are hydroxylamine, hydrazine, hydroxamic acids, oxime anions and the hypochlorite anion.

$$C_6H_5CH_2Br \xrightarrow{OH^-} C_6H_5CH_2OH + Br^-$$
$$C_6H_5CH_2Br \xrightarrow{OOH^-} C_6H_5CH_2OOH + Br^- \qquad [2.12]$$

The one feature common to all of these nucleophiles is the presence of an electronegative atom, containing one or more pairs of unshared electrons, adjacent to the nucleophilic atom. The increased reactivity shown by this class of nucleophiles has been called the α-effect (Edwards and Pearson, 1962). Whilst this effect is widely recognised, there is at present no generally accepted explanation of it. If it is postulated that in the displacement reaction a pair of electrons tends to leave the nucleophile, then the α-effect can be explained in terms of the stabilisation of the incipient positive charge on the nucleophilic atom by the electrons on the adjacent atom.

2.2.5. Solvation and ionic association. In the preceding sections nucleophilic reactivity has been discussed in terms of the structure and properties of the nucleophile, but other factors, such as solvation and ionic association, should also be considered. The effect of solvation will be discussed in more detail in a later section (§2.4.2), but one point should be made here. In protic solvents the nucleophilicity depends upon the size of the nucleophile rather than upon its basicity, and this has been explained in terms of the polarisability effect. It is well known that in protic solvents the smaller anions tend to be the more strongly solvated. Strong specific interactions with the solvent will reduce the reactivity of an anion, and so the higher reactivities of the larger nucleophiles, which have been discussed in terms of polarisability, might be due, in part at least, to changes in solvation in the direction of weaker interactions.

The nucleophilic order $Cl^- > Br^- > I^-$ has been determined from the reactions of tetra-n-butylammonium halides with n-butyl brosylate† and isobutyl tosylate in acetone (Winstein, Savedoff, Smith, Stevens and

† *p*–bromobenzenesulphonate, –OBs.

Gall, 1960). When lithium salts are used as the source of halide anion the nucleophilic order (based on observed reaction rates) is reversed. Conductivity measurements indicate that the lithium salts are present largely as ion pairs in acetone; the extent of ion-pair formation depending upon the ionic size. Thus the extent of ion pairing is greater with LiCl than with LiI, and is relatively unimportant with the tetra-n-butylammonium halides. The order of free anion reactivities for the lithium salts, obtained after correction for incomplete ionic dissociation, is $Cl^- > Br^- > I^-$. Ion pairing is not important in aqueous solutions and yet the observed nucleophilic order for the halide ions is $I^- > Br^- > Cl^-$, the order being the reverse of that observed in acetone. Since the polarisabilities of these anions are not likely to be very different in the two solvents, this change in order is most likely due to solvation effects (§2.4.3).

2.2.6. Ambident nucleophiles. Some nucleophiles have more than one nucleophilic atom which may function as the reaction site, e.g.

$$CN^- \quad NO_2^- \quad SO_3^{2-} \quad (C_6H_5O)_2.POS^-$$

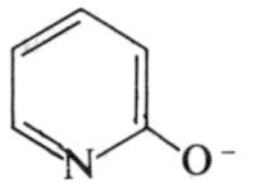

(2-pyridinolate anion: pyridine ring with O^- on the carbon adjacent to N)

Such reagents are called ambident nucleophiles. A typical example is the nitrite anion; silver nitrite reacts with alkyl halides to give mixtures of nitroalkanes and alkyl nitrites. The proportions vary with the experimental conditions in such a way as to suggest that the oxygen atom of the nucleophile is the preferred reaction site when the mechanism of the reaction is S_N1. This can be explained by assuming that the maximum electrostatic stabilisation of the transition state results from the interaction of the carbonium ion with the most electronegative centre in the nucleophile. When the mechanism becomes S_N2 other factors appear to be more important in determining the reaction site.

The sulphite, sulphinate and thiosulphate anions always form bonds to the sulphur atom in their reactions in protic solvents. The greater reactivity of the sulphur atom over the oxygen atom may be explained either by the larger polarisability of sulphur, or by a specific interaction of the solvent with the oxygen atom which decreases the reactivity of that centre.

2.2.7. Quantitive estimates of nucleophilic reactivity. Several approaches have been suggested for establishing a relative order of reactivity for nucleophiles, and the various methods have been reviewed (Ibne-Rasa, 1967). A convenient, and simple, measure of relative nucleophilic reactivity is provided by the rates at which a series of nucleophiles react with a given substrate. Swain and Scott (1953) used the rates of the S_N2 reactions of various nucleophiles with methyl bromide. They defined the nucleophilicity constant, n, by (2.4). In this

$$\log (k/k_0) = n \tag{2.4}$$

equation k is the second-order rate constant for the reaction of the nucleophile with methyl bromide and k_0 is the second-order rate constant for the hydrolysis of methyl bromide (calculated from the observed pseudo-first-order rate constant and the concentration of water in the solvent). Relative rates of reaction with other substrates may then be calculated by (2.5) in which s is the substrate constant which

TABLE 2.7 *Nucleophilic reactivity constants*

Nucleophile	n^a	$n_{CH_3I}{}^b$	Nucleophile	n	n_{CH_3I}
H_2O	0.00	–	OH^-	4.20	–
CH_3OH	–	0.00	CH_3O^-	–	6.29
NO_3^-	1.03	1.5	$(CH_3)_2Se$	–	6.32
$(C_6H_5)_3Sb$	–	2.0	NH_2OH	–	6.60
F^-	2.0	2.7	NH_2NH_2	–	6.61
SO_4^{2-}	2.5	3.5	$(C_2H_5)_3N$	–	6.66
CH_3COO^-	2.72	4.3	SCN^-	4.77	6.70
Cl^-	3.04	4.37	CN^-	5.1	6.70
$C_6H_5COO^-$	–	4.5	$(C_2H_5)_3As$	–	6.90
α-Picoline	–	4.7	$(C_2H_5)_2NH$	–	7.0
$(C_6H_5)_3As$	–	4.77	$(C_6H_5)_3P$	–	7.0
Imidazole	–	4.97	Pyrrolidine	–	7.23
$(CH_3O)_2P$	–	5.2	$SC(NH_2)_2$	4.1	7.27
Pyridine	3.6	5.23	Piperidine	–	7.30
HCO_3^-	3.8	–	I^-	5.04	7.42
HPO_4^-	3.8	–	HOO^-	–	7.8
$(C_2H_5)_2S$	–	5.34	$SeCN^-$	–	7.85
NO_2^-	–	5.35	HS^-	5.1	8
NH_3	–	5.50	$(C_6H_5)_3Pb^-$	–	8
$(CH_3)_2S$	–	5.54	SO_3^{2-}	5.1	8.53
$C_6H_5N(CH_3)_2$	–	5.64	$(C_2H_5)_3P$	–	8.72
C_6H_5SH	–	5·70	$S_2O_3^{2-}$	6.36	8.95
$C_6H_5NH_2$	4.49	5.70	$C_6H_5S^-$	–	9.92
$C_6H_5O^-$	–	5.75	$C_6H_5Se^-$	–	10.7
N_3^-	4.00	5.78	$(C_6H_5)_3Sn^-$	–	11.5
Br^-	3.89	5.79	$(C_6H_5)_3Ge^-$	–	12

[a] Water, 25 °C; Data from Ibne–Rasa (1967).
[b] Methanol, 25 °C; Data from Pearson, Sobel and Songstad (1968).

measures the sensitivity of the substrate towards the various nucleophiles. By definition, therefore, $n = 0.00$ for the nucleophile water and $s = 1.00$ for the substrate methyl bromide. Equation (2.5) enables

$$\log (k/k_0) = sn \tag{2.5}$$

relative reactivities to be calculated for a limited, but useful, range of experimental conditions. The n-values (table 2.7) provide a reliable guide to nucleophilic reactivity only for those displacements occurring at a saturated carbon atom in water or in aqueous-organic solvents. Serious deviations are observed when the nature of the reacting centre of the substrate changes, as, for example, when the electrophilic atom becomes sulphur instead of carbon. A disadvantage of the equation is that values of the substrate constant, s, have been evaluated for only a limited number of substrates (table 2.8).

TABLE 2.8 *Substrate constants*[a]

Substrate	s	Substrate	s
Ethyl tosylate	0.66	Methyl bromide	1.00
Benzyl chloride	0.87	β-Propiolactone	0.77
Epichlorohydrin	0.93	Benzoyl chloride	1.43
Glycidol	1.00	Benzene sulphonyl chloride	1.25
Mustard cation	0.95		

[a] Data from Swain and Scott (1953).

A more recent study of nucleophilic reactivity has provided data from which n-values may be obtained for an extended range of nucleophiles (Pearson, Sobel and Songstad, 1968). An approach similar to that of Swain and Scott was employed; n_{CH_3I}-values were calculated by (2.4) using methyl iodide as the standard substrate and methanol as the standard nucleophile. The n_{CH_3I}-values (table 2.7) are linearly related to the n-values of Swain and Scott (fig. 2.7) and the slope of the correlation lines gives a proportionality constant of 1.4, (2.6).

$$n_{CH3I} = 1.4n \tag{2.6}$$

The data upon which the above nucleophilicity scales are based come from reactions in protic solvents (water, alcohols and mixed aqueous–organic solvents); nucleophilic reactivities will be successfully predicted by (2.5) only for reactions in such solvents. The change to an aprotic solvent can have a profound effect on nucleophilic reactivity, reversing many of the trends observed in protic solvents. These changes cannot

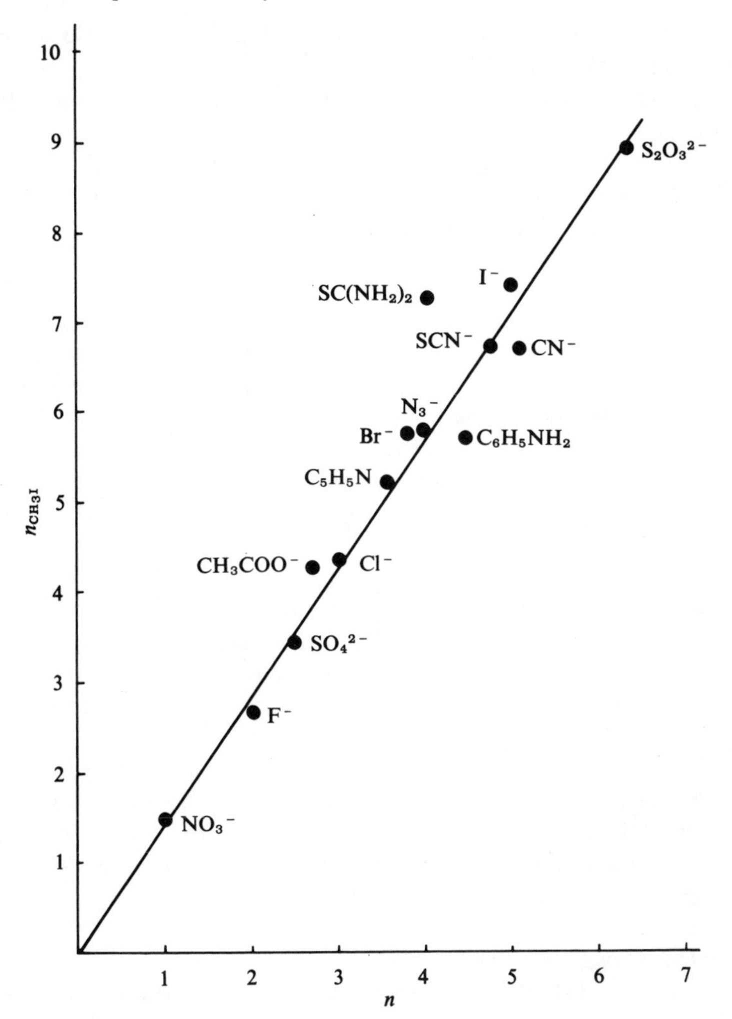

Fig. 2.7 Correlation between the nucleophilic reactivity constants n and n_{CH_3I} (cf. table 2.7).

be taken into account by varying the substrate constant, and so we should not expect to be able to use (2.5), in an unmodified form, to calculate reactivities in both protic and aprotic solvents with only a single set of n-values. The effect of solvent on nucleophilicity can be accommodated by (2.5) by the inclusion of an additional term which takes into account the solvent activity coefficients (see p. 54) of the nucleophiles (Parker, 1969).

Another correlation was suggested by Edwards (1956), in which the nucleophilic reactivity was related to two parameters; one a measure of the basicity and the other a measure of the polarisability of the nucleophile. In one form the relation was written as (2.7), in which k and k_0 are rate constants for the reaction of the substrate with the nucleophile

$$\log(k/k_0) = AP + BH \tag{2.7}$$

and with water. The parameter H is a measure of the basicity of the nucleophile and is obtained from the $\mathrm{p}K_a$, after correction for the $\mathrm{p}K_a$ of H_3O^+ (1.74); thus by definition, $H = \mathrm{p}K_a + 1.74$. P is a measure of the polarisability of the nucleophile and is defined as $P \equiv \log(R/R_{H_2O})$, where the R terms are molar refractions. The remaining terms in (2.7), A and B, are substrate parameters which measure the sensitivity of the substrate to changes in polarisability and basicity, respectively, of the nucleophile.

One advantage of the two-parameter equation is that it allows for a variation in nucleophilic reactivity. The relative importance of the two properties of the nucleophile in determining its reactivity may change from one type of substrate to another; this is allowed for by a variation in the substrate constants A and B. The Edwards equation thus has greater flexibility and applies over a wider range of conditions than does the Swain–Scott equation, although it is necessary to evaluate a larger number of parameters. It might be inferred from the success of (2.7) that the basicity and polarisability of the nucleophile are the only factors of importance in determining its reactivity; however, such an interpretation should be used with caution.

2.3. The leaving ability of groups

The rate of a substitution reaction occurring either by the unimolecular or by the bimolecular mechanism is sensitive to the nature of the leaving group. In both cases formation of the transition state requires a stretching of the bond between the carbon atom and the leaving group, and so will be facilitated when that bond is relatively weak. Some transfer of charge also takes place towards the displaced group and the reaction will therefore be favoured by an increase in the ability of the leaving group to bear a negative charge. The latter circumstance implies that the leaving ability of an atom or group should be related to its basicity; the less basic groups being the more easily displaced. For structurally similar leaving groups, linear correlations have been observed between the rate of reaction and the basicity of the leaving group. However, since a leaving group is essentially a nucleophile departing from the reaction centre, we might expect, from the discussion of the previous section, that properties other than basicity are important in determining the leaving ability.

One such factor is the nature of the interaction between the leaving group and the solvent. Hydrogen bonding can assist the departure of the leaving group and thus enhance the leaving ability. A more extreme type of interaction is protonation which, because it reduces the basicity of the leaving group, will also result in an increase in the ease of displacement. This effect is well known from the reactions of alcohols and ethers; whereas —OH and —OR are not readily displaced from a saturated carbon atom, the protonated forms, $—OH_2^+$ and $—ORH^+$, are good leaving groups. The importance of the solvent interactions in assisting the departure of the leaving group will in general vary among the different solvents and the different leaving groups. Therefore, a given order of leaving abilities should be dependent on the nature of the solvent (§2.4.3). The order of leaving abilities of the more common groups frequently observed in protic solvents is the following

$$Br-C_6H_4-O_2SO^- > CH_3-C_6H_4-O_2SO^- > I^- > Br^- > Cl^- >$$

$$F^- > CH_3COO^- > \overset{+}{N}R_3 > OR > OH > NR_2$$

Little progress has been made towards a quantitative understanding of leaving abilities, although two approaches to the problem have been suggested (Davis, 1965). In the first, relative reactivities for the standard reaction [*2.13*] in methanol at 25 °C (Br^- is the standard leaving group) are used to define a set of leaving group constants. The relatives rates of

$$CH_3O^- + CH_3X \longrightarrow CH_3OCH_3 + X^- \qquad [2.13]$$

substitution with other substrates are then calculated by an equation containing two parameters which is formally similar to the Swain and Scott equation. In the second approach, use is made of the Edwards equation; however, neither method appears to have been widely used.

2.4. Solvent effects

The effect of a change of solvent on the rate constant of reaction [*2.14*] may be expressed in terms of the effect of the change of solvent on the

Gibbs free energy of activation (2.8). In this equation and those which

$$\mathrm{N + RX \rightleftharpoons [NRX]^{\ddagger} \longrightarrow product} \qquad [2.14]$$

$$\ln(k_s/k_0) = (-1/RT)(\Delta G_s^{\ddagger} - \Delta G_0^{\ddagger}) \qquad (2.8)$$

follow, s refers to the solvent of interest and 0 to the reference solvent. The Gibbs free energy of activation is equal to the change in the standard chemical potential accompanying formation of the transition state, i.e. $\Delta G^{\ddagger} = \Delta\mu^{\ddagger} = (\mu_{\ddagger} - \mu_N - \mu_{RX})$. We may therefore write (2.8) in the form of (2.9). The terms on the right-hand side of (2.9) represent the

$$\ln(k_s/k_0) = (-1/RT)\{(\mu_{\ddagger}^{s} - \mu_{\ddagger}^{0}) - (\mu_N^{s} - \mu_N^{0}) - (\mu_{RX}^{s} - \mu_{RX}^{0})\} \qquad (2.9)$$

changes in the standard chemical potentials of the species involved in the reaction on transfer from the reference solvent to the solvent of interest. These terms are related to the solvent activity coefficients (standard activities) by (2.10). Using this result, we may rewrite (2.9)

$$\ln {}^{0}\gamma_A^{s} = (1/RT)(\mu_A^{s} - \mu_A^{0}) \qquad (2.10)$$

in the desired form, (2.11). The effect of solvent on the rate of reaction is

$$\ln(k_s/k_0) = \ln {}^{0}\gamma_N^{s} + \ln {}^{0}\gamma_{RX}^{s} - \ln {}^{0}\gamma_{\ddagger}^{s} \qquad (2.11)$$

therefore given in terms of a sum of the appropriate solvent activity coefficients. Values of ${}^{0}\gamma^{s}$ for the reactants may be obtained from measurements of solubilities, Henry's law constants, or distribution coefficients, and they represent the changes in the solvation of the reactants accompanying solvent transfer. The values of the corresponding transition state terms may be calculated from (2.11). A positive value of ${}^{0}\gamma^{s}$ implies that the solute is better solvated in the reference solvent than in the solvent of interest. For charged species, single-ion solvent activity coefficients may be obtained by making use of one of the standard extrathermodynamic assumptions, for example that ${}^{0}\gamma^{s}{}_{Ph_4As^+} = {}^{0}\gamma^{s}{}_{Ph_4B^-}$ (Parker, 1969).

The solvation of a solute is usually considered to involve electrostatic ion–dipole, or dipole–dipole, interactions on which may be superimposed more specific types of interactions, such as the formation of strong hydrogen bonds. Solvent–solvent interactions might also be of importance, since the formation of solvent–solute interactions leads to the orientation of the solvent about the solute, a process which causes disruption of any existing solvent structure.

The rates of many S_N2 reactions increase on changing from a protic to a dipolar aprotic solvent (table 2.9). The primary cause appears to be a large change in the solvation of the anionic nucleophile. Protic solvents, e.g. methanol, are hydrogen bond donors and good solvators for anions. Dipolar aprotic solvents, e.g. dimethylformamide (DMF),

TABLE 2.9 *The effect of changing the solvent from methanol to dimethylformamide on the rates of some S_N2 reactions at 25 °C*[a]

Reactants	$\log (k^D/k^M)$	$\log {}^M\gamma_{RX}{}^D$	$\log {}^M\gamma_N{}^D$	$\log {}^M\gamma_{\ddagger}{}^D$
$CH_3Cl + N_3^-$	3.3	−0.4	4.9	1.2
$CH_3Cl + SCN^-$	1.4	−0.4	2.7	0.9
$CH_3Br + N_3^-$	3.9	−0.3	4.9	0.7
$CH_3Br + SCN^-$	1.7	−0.3	2.7	0.7
$CH_3I + N_3^-$	4.6	−0.5	4.9	−0.2
$CH_3I + SCN^-$	2.2	−0.5	2.7	0.0
$CH_3I + OAc^-$	6.9	−0.5	9.2	1.8
$CH_3OTs + N_3^-$	2.0	−0.6	4.9	2.3
$CH_3OTs + SCN^-$	0.8	−0.6	2.7	1.3
$n\text{-}BuBr + N_3^-$	3.4	0.0	4.9	1.5
$i\text{-}PrBr + N_3^-$	2.7	−0.1	4.9	2.1
$C_6H_5CH_2Cl + N_3^-$	2.4	0.2	4.9	2.7
$CH_3\overset{+}{S}Me_2 + N_3^-$	3.1	−3.1	4.9	−1.3
$n\text{-}BuBr + C_5H_5N$	0.8	−0.1	C[b]	−0.9 + C[b]

[a] Data from Parker (1969). [b] $\log {}^M\gamma_{C_5H_5N}{}^D$ unknown.

on the other hand, cannot form strong hydrogen bonding interactions, and anions appear to be poorly solvated in these solvents. The much smaller changes in the solvation of the transition states on solvent transfer are consistent with either weaker interactions in methanol, or stronger interactions in DMF than are found with the corresponding reactant anions. The last two entries in table 2.9 are for reactions of different charge types. With a positively charged substrate a rate increase is observed on going from methanol to DMF, but the effect of decreased solvation of the anion is counterbalanced to some extent by increased solvation of the cationic substrate. Reactions involving uncharged nucleophiles are not greatly affected by the change from a protic to a dipolar aprotic solvent.

Reactions occurring by the S_N1 mechanism proceed most readily in protic solvents, since both the cation and the counter ion are effectively solvated. Qualitative discussions of solvent effects usually refer to the 'polarity' of the solvent, a term used to describe the ability of the solvent to solvate polar molecules and reaction intermediates; ions and dipolar molecules are better solvated by a more polar solvent. Polarity is not related in any simple way to the properties of the solvent, e.g. the dielectric constant. A number of empirical measures of solvent polarity have been obtained, however, from the effect of the solvent on various standard reactions (cf. Reichardt, 1965). The Winstein–Grunwald scale of *Y*-values is one such measure, convenient for predicting the effect of solvent changes on the rates of solvolysis reactions (§3.6.3).

2.4.1. The Hughes–Ingold theory of solvent effects. This accounts for the effect of a change in the polarity of the solvent on the rates of S_N reactions in terms of the solvation of the reactants and the transition states (Ingold, 1969). It assumes that for a given solvent the degree of solvation will:

(i) increase with the magnitude of the charge on the solute,
(ii) decrease with increased distribution of a given charge.

Thus a reaction in which the formation of the transition state involves an increase in net charge, or the concentration of a given charge, should be faster in a more polar solvent. On the other hand, a reaction in which formation of the transition state involves a decrease in net charge, or the dispersion of a given charge, should be slower in a more polar solvent. Furthermore, the effects are expected to be larger when charges are created or destroyed than when they are merely concentrated or dispersed (table 2.10).

TABLE 2.10 *Solvent effects predicted for S_N reactions of different charge-types by the Hughes–Ingold theory*

Mechanism	Reactants	Transition state	Charge in transition state relative to reactants	Effect of increase in polarity of solvent on rate
S_N2	$Y^- + RX$	$^{\delta-}Y\cdots R\cdots X^{\delta-}$	Dispersed	Small decrease
S_N2	$Y + RX$	$^{\delta+}Y\cdots R\cdots X^{\delta-}$	Increased	Large increase
S_N2	$Y^- + RX^+$	$^{\delta-}Y\cdots R\cdots X^{\delta+}$	Decreased	Large decrease
S_N2	$Y + RX^+$	$^{\delta+}Y\cdots R\cdots X^{\delta+}$	Dispersed	Small decrease
S_N1	RX	$^{\delta+}R\cdots X^{\delta-}$	Increased	Large increase
S_N1	RX^+	$^{\delta+}R\cdots X^{\delta+}$	Dispersed	Small decrease

The theory has correctly predicted solvent effects on many types of reactions, although in some cases the success has been quite fortuitous

(Bunton, 1963). The changes in solvation are considered to modify only the enthalpy of activation, it being assumed that changes in the entropy of activation are unimportant. Reliable predictions of solvent effects can only be expected therefore for reactions in which the Gibbs free energy of activation is determined largely by the enthalpy term (cf. §3.5).

2.4.2. The effect of solvent on nucleophilicity. The change from a protic to a dipolar aprotic solvent can not only lead to large changes in the rates of S_N2 reactions, but also it can lead to a change in the order of relative reactivity of a series of nucleophiles towards a given substrate. The relative rates, based on that of thiocyanate ion, for the bimolecular reactions of various nucleophiles and methyl iodide, in the solvents methanol and DMF, are given in table 2.11. The effect of the solvent change on the relative rates is larger for the more compact anions such as chloride, bromide or azide than for the larger anions such as iodide.

TABLE 2.11 *Relative rates at 25 °C for the bimolecular substitutions*[a] $N^- + CH_3I \rightarrow CH_3N + I^-$

Nucleophile (N^-)	$\log (k_{N^-}/k_{SCN^-})$	
	Methanol	DMF
$ClCH_2COO^-$	−3.4	+0.5
CH_3COO^-	−2.4	+2.4
$p\text{-}NO_2.C_6H_4O^-$	−2.2	−0.8
Cl^-	−2.33	+1.5
Br^-	−0.91	+1.2
N_3^-	−0.92	+1.6
SCN^-	0.00	0.0
CN^-	0.00	+3.6
I^-	+0.7	+0.7
$p\text{-}NO_2.C_6H_4S^-$	+2.1	+2.3
$C_6H_5.S^-$	+3.2	+5·0

[a] Data from Parker (1969).

The changes observed are consistent with the ideas (see p. 55) that anions are more solvated in methanol than in DMF and that the degree of solvation in the former solvent depends upon the nature of the anion. Size is one factor and in general smaller anions are better solvated than larger ones. Structural features in the anion which lead to strong hydrogen bonding interactions with the solvent also increase solvation. Thus the largest changes in relative rates (table 2.11) are observed for those anions which are expected to be particularly well solvated in methanol; the reactivity of acetate ion, relative to that of thiocyanate ion, for example, changes by a factor of *ca* 10^5.

The nucleophilic order observed in methanol, $I^- > SCN^-, CN^- > N_3^-, Br^- > Cl^- > CH_3COO^-$ is that commonly found in protic solvents (cf. table 2.7) and is the order usually quoted for relative nucleophilicities towards a saturated carbon atom (§2.2.2). However, this order is not maintained in dipolar aprotic solvents owing to the differential solvent effects observed with the different anions on transfer from a protic solvent. Thus for the above anions in DMF the nucleophilic order becomes, $CN^- > CH_3COO^- > Cl^-$, Br, $N_3^- > I^- > SCN^-$. This sensitivity of the nucleophilicities of anions to the solvent indicates that the reactivity of an anion towards a given substrate is determined, in part at least, by the solvation of the anion.

2.4.3. The effect of solvent on leaving ability. The importance of solvation in determining reactivity, as noted above for reactant anions, extends also to leaving groups. The effect on the relative leaving ability of a group of a change from a protic to a dipolar aprotic solvent can be explained in terms of the stronger interactions between protic solvents and the leaving group, particularly when the latter is a small anion. Thus methanol levels the leaving tendencies of the halide ions (relative to their values in DMF), because it tends to favour the departure of the smaller chloride ion rather than the larger iodide ion. It has been suggested that the behaviour in dipolar aprotic solvents, in which interactions between the leaving group and the solvent are minimal, should give a better indication of intrinsic leaving ability (Parker, 1969).

3 Characterisation of mechanism

3.1. Kinetics of nucleophilic substitutions

A classification of nucleophilic substitutions was given in chapter 1, but without any discussion of the kinetic forms associated with the different mechanisms. It is perhaps worth emphasising that the numbers in the terms S_N1 and S_N2 refer to the molecularities of the substitution reactions and not to the observed kinetic orders of these processes. Since kinetic studies generally provide one of the most useful means of probing reaction mechanisms, it is important to understand the relationships between the kinetics and mechanisms of nucleophilic substitutions.

The S_N2 reaction involves a single step in which two molecules are undergoing a covalency change, and the transition state is formally composed of one molecule of the substrate and one molecule of the nucleophile. The kinetic expression for such a situation is that given in (3.1).

$$-\,d[RX]/dt = k_2[RX][N] \qquad (3.1)$$

Second-order kinetics will be observed, and the rate of substitution will be proportional to the concentrations of both RX and N, if comparable amounts of each reactant are present. If, however, a large excess of the nucleophile is employed, or if the solvent acts as the nucleophile, then the concentration of N will remain effectively constant throughout the reaction and first-order kinetics will be observed (3.2). This latter case

$$-\,d[RX]/dt = k_1'[RX] \qquad (3.2)$$

always obtains for solvolytic displacements, and for these reactions the observed kinetic order is no criterion of mechanism. The nucleophilic species present in the transition state need not be the added nucleophile, but could be formed from it in a pre-equilibrium. In such a case, first-order kinetics could again be observed. For reactions in solvents of low dielectric constant ion pairing of a charged nucleophile might become important and this will also affect the observed kinetic order.

An S_N1 mechanism which involves the irreversible formation of a carbonium ion in the rate-limiting step, would give first-order kinetics, (3.3).

$$-\,d[RX]/dt = k_1[RX] \qquad (3.3)$$

A more general representation of the S_N1 mechanism allows for reversal of the initial heterolysis, as shown in [*3.1*]. Application of the

$$RX \underset{k_{-1}}{\overset{k_1}{\rightleftharpoons}} R^+ + X^-$$
$$R^+ + N \xrightarrow{k_N} RN^+ \qquad [3.1]$$

steady state approximation to this mechanism leads to the following kinetic expression (3.4). For simplicity it is assumed that the carbonium

$$-\,d[RX]/dt = k_1[RX]/\{1 + (k_{-1}[X^-]/k_N[N])\} \qquad (3.4)$$

ion reacts with only one nucleophile, but the kinetic expression is easily modified to allow for reactions with more than one nucleophile by replacing the term $k_N[N]$ by the summation $\sum_i k_{N_i}[N_i]$. Depending on the magnitudes of the terms in the denominator of (3.4) several cases may be distinguished.

If $(k_{-1}[X^-]/k_N[N]) \ll 1$ throughout the reaction, then (3.4) reduces to (3.3) and first-order kinetics will be observed. For this condition to hold, the nucleophile, N, must be more efficient at trapping the carbonium ion than is the common ion X^-.

If $(k_{-1}[X^-]/k_N[N]) \gg 1$ throughout the reaction (3.4) simplifies to (3.5). If the concentration $[X^-]$ remains effectively constant, then second-

$$-\,d[RX]/dt = (k_1k_N/k_{-1}[X^-])[RX][N] \qquad (3.5)$$

order kinetics will be observed. This situation, in which consumption of the carbonium ion as well as its formation is kinetically important, corresponds to the $S_N2(C^+)$ mechanism (§1.4.1).

The kinetic expression (3.5), includes the rate constants k_1, k_{-1} and k_N, and so the overall rate of the reaction is determined by the rates of all the individual steps. It is useful when discussing mechanisms involving more than one step to be able to refer to the last step whose rate constant appears in the kinetic equation (Zollinger, 1964). The term rate-limiting is convenient for this purpose, and in the present example this would be the step with rate constant k_N.†

If the magnitude of $(k_{-1}[X^-]/k_N[N])$ varies during the course of the reaction, then a reaction which initially exhibits first-order kinetics will, as reaction proceeds, show deviations from simple first-order behaviour. This is usually an indication that the common ion X^- is becoming increasingly effective at reversing the initial heterolysis in [*3.1*] as the reaction progresses. This behaviour, when observable, is a diagnostic feature of the S_N1 mechanism (§3.2.1).

3.2. Salt effects

The effect of an added salt on the rate of a nucleophilic substitution can be expressed in terms of its effect on the activity coefficients of the reactants and the transition state using the Brønsted–Bjerrum rate equation (3.6). In this equation γ_N and γ_{RX} are the activity coefficients of

$$k = k_0(\gamma_N \gamma_{RX}/\gamma_{\ddagger}) \qquad (3.6)$$

the reactants and $\gamma_{\ddagger}$ is the activity coefficient of the transition state; k_0 and k are the rate constants in the absence and presence respectively of the salt. This approach requires that a relationship be established between the concentration of the added salt, or rather the ionic strength of the solution, and the various activity coefficients.

The Debye–Hückel theory provides such a relationship for ions in dilute solutions, but this theory is not suitable for the types of polar transition states usually thought to be involved in nucleophilic substitutions. A better model for these structures is provided by considering the ionic distribution around a dipole, for which the activity coefficient may be related to the ionic strength, I, and the dielectric constant, D, of the solution by (3.7) (Ingold, 1969). In this equation z represents the

$$\log \gamma \propto -\, z^2 dI/(DT)^2 \qquad (3.7)$$

fraction of charge at each point pole, d is their separation, and T is the absolute temperature.

The above approach assumes that the effect of added salts on the reaction rate arises from a non-specific ionic-atmosphere effect; the distribution of ions around the dipole provides an ionic-atmosphere which stabilises that dipole. The effect will be greater the larger the dipole, and it can be expected that the addition of salt will generally accelerate the

† The rate-limiting step has been defined in several different ways and this has necessarily led to some ambiguity; the above definition appears to be satisfactory.

rate of an S_N1 reaction, in which the transition state is more polar than the reactants, and have little effect on the rate of an S_N2 reaction in which charge is dispersed in the transition state. In addition, the form of (3.7) leads to two predictions concerning the nature of salt effects. First, they should be relatively more important in solvents of low dielectric constant. Secondly, the effect should not depend upon the actual identity of the salt, although it does depend upon its concentration; the logarithm of the rate constant being linearly related to the ionic strength.

This simple electrostatic model has been found to predict correctly the observed salt effects in the solvolysis of several alkyl halides, assuming that the change in ionic strength modifies only the activity coefficients of the more polar transition states (Ingold, 1969). This success indicates that the ionic-atmosphere effect is an important factor in determining the magnitude of salt effects, although several of the assumptions involved might be criticised. Thus, salt effects on the activity coefficients of the neutral substrates are ignored, even though it is known that salts show specific effects on the activity coefficients of non-electrolytes in water. It is also assumed that the added salt is fully dissociated, a satisfactory assumption for polar solvents of high dielectric constant, but not for solvents of low dielectric constant. An additional factor should also be considered when mixed solvents are used. This is a salt-induced medium effect, which can be regarded as a change in the effective composition of the solvent brought about by the specific solvation of the salt by the more polar component of the solvent (Jackson and Kohnstam, 1965). In aqueous organic solvents, for example, the salt might tend to 'dry' the solvent by attracting about itself water molecules into a tightly held solvation shell.

The effects of added salts on the hydrolysis of t-butyl chloride in water have been studied in detail, and the results show how specific salt effects might remain undetected (Clark and Taft, 1962). Measurements of the initial vapour pressure of the dissolved t-butyl chloride in the presence and absence of salts gave the salt effect on the activity coefficient of the substrate, γ_{RX}. By measuring the rates of hydrolysis in the presence and absence of salts, with a known pressure of the reactant above the solution (thereby keeping the activity of dissolved t-butyl chloride constant), it was possible to obtain the salt effects on the activity coefficient of the transition state, $\gamma_{\ddagger}$. Specific salt effects were observed on both γ_{RX} and $\gamma_{\ddagger}$, but these effects were almost identical for some of the salts studied and cancelled out when the ratio of the activity coefficients was taken.

In such cases the salt effect on the rate of hydrolysis was correctly predicted by the simple electrostatic model.

Salt effects have been determined for many reactions in solvents of low dielectric constant, and the results are in general not accounted for by the ionic-atmosphere model. Solvolysis reactions in acetic acid have been particularly well studied, and in this solvent the 'normal' salt effect can be described by the empirical relationship (3.8). This relationship is

$$k = k_0(1 + b[\text{salt}]) \tag{3.8}$$

usually valid for salt concentrations of up to *ca* 0.1 mol l^{-1}, k and k_0 are the rate constants in the presence and absence of salt, [salt] is the stoichiometric concentration of the salt in mol l^{-1}, and b is the salt parameter, being the slope of the graph of k against [salt]. The value of b therefore provides a convenient measure of the magnitude of the normal salt effect.

The effect of lithium perchlorate on the rate of ionisation of *p*-methoxyneophyl tosylate (47) is dependent upon the solvent (table 3.1). The

TABLE 3.1 *The normal salt effect of* $LiClO_4$ *on the rate of ionisation of p-methoxyneophyl tosylate* [a]

Solvent	Temperature (°C)	b	D
Dimethylsulphoxide	75	0.0	48.9
Dimethylformamide	75	1.4	36.7
Acetic acid	50	12.2	6.2
Acetone	75	47	20.5
Tetrahydrofuran	75	482	7.4
Ether	50	2.98×10^5	4.2

[a] Data from Winstein, Smith and Darwish (1959).

$CH_3O-C_6H_4-C(CH_3)_2-CH_2OTs$

(47)

magnitude of the salt effect (given by the salt parameter, b) increases with decreasing dielectric constant of the solvent, as required by the electrostatic treatment, but the position of acetic acid indicates that the dielectric constant is not the only factor which determines the salt effect. Another difficulty is that salt effects also depend upon the nature of the salt; thus in the ionisation of (47) in acetic acid at 50 °C, the values of the salt parameter for the salts $LiClO_4$, LiOTs and NaOAc are 12.2, 2.5 and 0.5 respectively. Such specific effects are not predicted by the electrostatic treatment based on ion–dipole interactions (3.7).

In addition to varying with the solvent and the nature of the salt, the value of b also varies with the nature of the substrate and with the temperature. This behaviour of the normal salt effect can be rationalised in terms of an electrostatic model based on dipole–dipole interactions between the salt (present as an ion pair) and the dipolar transition state (Perrin and Pressing, 1971). Apparently the association into ion pairs is an important factor contributing to the behaviour of added salts on the rates of reactions in media of low dielectric constant.

3.2.1. The common-ion effect. The specific rates of hydrolysis of benzhydryl chloride and bromide in aqueous acetone are found to decrease steadily as the reactions progress. Moreover, the addition of a salt having the same anion as that produced by the heterolysis of the substrate is found to depress the rate of solvolysis (table 3.2). This effect of the common-ion salt is in marked contrast to the usual type of behaviour observed with added salts. Similar effects have been observed with other substrates; for instance, the rate of hydrolysis of triphenylmethyl chloride in aqueous acetone is decreased fourfold by sodium chloride (0.01 mol l^{-1}).

TABLE 3.2 *The effect of salts on the rates of solvolysis of benzhydryl halides in 80 per cent aqueous acetone at 25 °C*[a]

Salt	Concentration (mol l^{-1})	$10^5 k_1$ (s^{-1})	
		$(C_6H_5)_2CHBr$	$(C_6H_5)_2CHCl$
–	–	153	7.00
LiBr	0.1	133	8.16
LiCl	0.1	194	6.09

[a] Data from Benfey, Hughes and Ingold (1952).

For an S_N1 solvolysis (3.4) may be expressed as (3.9) in which the constant k_N' has replaced the term k_N[N], since the solvent is now the nucleophile and its concentration remains effectively constant. If the

$$-\,d[RX]/dt = k_1[RX]/\{1 + (k_{-1}[X^-]/k_N')\} \qquad (3.9)$$

leaving anion X^- cannot compete with the solvent for the carbonium ion, then $k_{-1}[X^-] \ll k_N'$ and (3.9) reduces to the form of a simple first-order reaction. If, however, the term $k_{-1}[X^-]/k_N'$ is significant at the start of the reaction it will become progressively more important as the reaction proceeds, because the concentration of X^- gradually increases. The rate of reaction will therefore become gradually smaller (3.9). The importance of the term $k_{-1}[X^-]/k_N'$ may also be increased by the addition of salt with the common anion X^-. This decrease in the rate of solvolysis due to the presence of the leaving anion is called the common-ion or mass-law effect. It should be noted that the decrease in the rate of solvolysis is not caused by a reduction in the rate of ionisation (i.e. $k_1[RX]$), but by the capture of the carbonium ion by the common ion X^- to re-form starting material; the rate of ionisation is, in fact, subject to a normal salt effect.

The types of substrates for which the common-ion effect has been observed all have structures which might be expected to produce relatively stable carbonium ions. Indeed, it is usually assumed that the stability of the carbonium ion determines whether the term $k_{-1}[X^-]/k_N'$ will be important or not. The explanation commonly given is that the more stable, and hence the longer lived, carbonium ion is more selective towards the available nucleophiles. A stable carbonium ion may be able to discriminate between the more reactive ion X^-, present in low concentration, and the less reactive solvent.

3.2.2. Trapping of the carbonium ion. In solvolyses of benzhydryl halides in aqueous acetone, the presence of low concentrations of sodium azide causes small accelerations in reaction rate ($-d[RX]/dt$), consistent with a normal salt effect. However, it is found that the sodium azide is effective in diverting some of the reaction from solvolysis to the production of azide. The amount of azide in the products appears to be independent of the identity of the halide leaving group. In 90 per cent aqueous acetone benzhydryl bromide reacts just over 33 times faster than does benzhydryl chloride in the presence of sodium azide (0.1 mol l^{-1}), but the percentage of product diverted into azide is about the same (bromide 33.5 per cent, chloride 34 per cent). These results seem to be best explained in terms of a mechanism which involves the trapping of an intermediate of the solvolysis reaction by the added salt [*3.2*]. The amount of alkyl azide produced depends upon how effectively N_3^- competes with the other nucleophiles present for the carbonium ion (or other solvolysis intermediate). Since this product-determining step is not related to the slow rate-determining step, which in this case is the initial ionisation, there will be no simple relationship between the effect of added azide ion on the rate of reaction and the amount of alkyl azide produced.

$$\begin{array}{ccccc} RX & \rightleftharpoons & R^+ + X^- & \xrightarrow{H_2O} & ROH + H^+ + X^- \\ & & \downarrow{N_3^-} & & \\ & & RN_3 + X^- & & \end{array} \qquad [3.2]$$

Under certain conditions the accelerated rate of reaction observed on the addition of sodium azide might be the result of an increased rate of destruction of the substrate due to the additional S_N2 pathway [*3.3*] being

$$N_3^- + RX \longrightarrow RN_3 + X^- \qquad [3.3]$$

provided. In such cases the rate-determining step for production of azide is the same as the product-determining step, and it might be expected that a simple relationship should exist between the increase in the rate of reaction and the amount of alkyl azide produced (however, cf. §4.3.4).

3.3. Stereochemistry of nucleophilic substitutions

Some of the stereochemical consequences of the S_N1 and S_N2 mechanisms were mentioned in chapter 1, but they were not commented upon there. The stereochemistry of the S_N2 mechanism is quite characteristic; inversion of configuration at the reacting carbon atom is always observed (Walden Inversion). The stereochemistry of the S_N1 mechanism, on the other hand, is more variable, any result from complete inversion to complete retention of configuration being possible.

3.3.1. S_N2 reactions. The bimolecular mechanism of substitution was shown to proceed with inversion of configuration for the reaction between radio-iodide ion and optically active 2-octyl iodide in dry acetone [*3.4*], by comparing the rate of radio-iodide exchange with the rate of change of optical activity (Hughes, Juliusberger, Masterman, Topley and Weiss, 1935). The substitution was found to proceed with second-

$$^*I^- + (C_6H_{13})(H)(CH_3)C{-}I \longrightarrow {}^*I{-}C(C_6H_{13})(H)(CH_3) + I^- \qquad [3.4]$$

order kinetics and therefore presumably by the S_N2 mechanism. The rate of change of optical activity was found to be equal to the rate of isotopic exchange, which implies that each act of substitution occurs with inversion of configuration. This requires that the nucleophile attack the carbon atom from the rear (48) rather than on the front side (49).

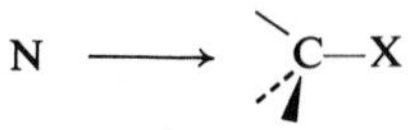

N ⟶ C—X

(48)

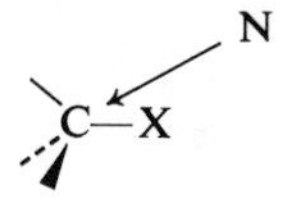

N ↘ C—X

(49)

One explanation of this preferred direction of attack assumed that the electrostatic repulsion between the incoming and leaving groups was the factor responsible. However, this now appears to be unlikely, because the stereochemistry has been shown to be independent of the charges associated with N and X. The general method has been to correlate the stereochemistry of S_N2 reactions of the different charge types (§1.1) with that for the type illustrated in [*3.4*], which has been shown to occur with inversion. The method of correlation is illustrated by scheme [*3.5*].

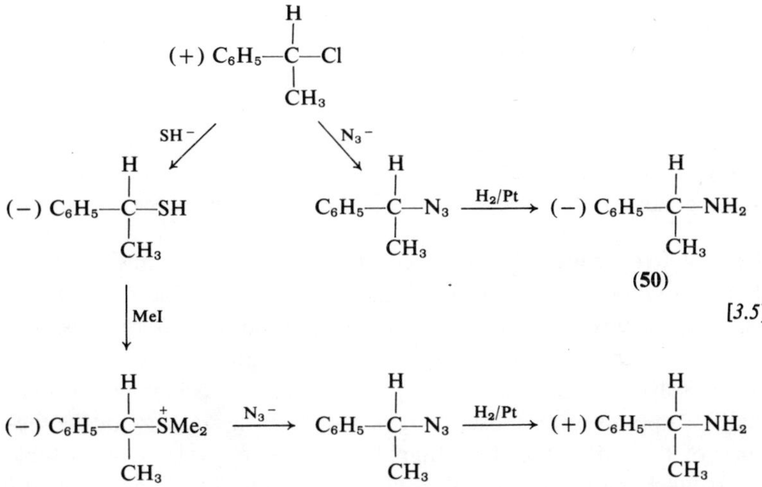

$$(+)\ C_6H_5\text{—}CH(CH_3)\text{—}Cl$$

$$\xrightarrow{SH^-} (-)\ C_6H_5\text{—}CH(CH_3)\text{—}SH$$

$$\xrightarrow{N_3^-} C_6H_5\text{—}CH(CH_3)\text{—}N_3 \xrightarrow{H_2/Pt} (-)\ C_6H_5\text{—}CH(CH_3)\text{—}NH_2 \quad \textbf{(50)}$$

$$(-)\ C_6H_5\text{—}CH(CH_3)\text{—}SH \xrightarrow{MeI} (-)\ C_6H_5\text{—}CH(CH_3)\text{—}\overset{+}{S}Me_2 \xrightarrow{N_3^-} C_6H_5\text{—}CH(CH_3)\text{—}N_3 \xrightarrow{H_2/Pt} (+)\ C_6H_5\text{—}CH(CH_3)\text{—}NH_2 \quad \textbf{(51)} \qquad [3.5]$$

The S_N2 reaction of 1-phenylethyl chloride and azide ion gives an azide which on reduction (which does not touch the asymmetric carbon atom) yields the amine (50). The same 1-phenylethyl chloride is converted by an S_N2 reaction into a thiol which can then be methylated (again without touching the asymmetric carbon atom). Since the S_N2

reactions of both N_3^- and SH^- with 1-phenylethyl chloride are of the type which has been shown to proceed with inversion, the amine (50) and the sulphonium ion are known to be of like configuration. The sulphonium ion can now be converted into an azide by an S_N2 reaction, and the azide reduced to give the amine (51), which is the enantiomer of (50). This result shows that the S_N2 reaction between the sulphonium ion and the azide ion must have taken place with inversion of configuration. Thus even when the leaving group is positively charged and the nucleophile is negatively charged the S_N2 reaction proceeds with inversion (Ingold, 1969).

The observation that S_N2 reactions at carbon always give inversion of configuration can be rationalised in terms of MO theory (Gilchrist and Storr, 1972). One argument, based on Frontier MOs, assumes that the direction of attack of the nucleophile is controlled by the ease with which the highest occupied molecular orbital of the nucleophile can interact with the σ^* antibonding orbital of the C–X bond [*3.6*]. This interaction can occur most readily when the nucleophile approaches the

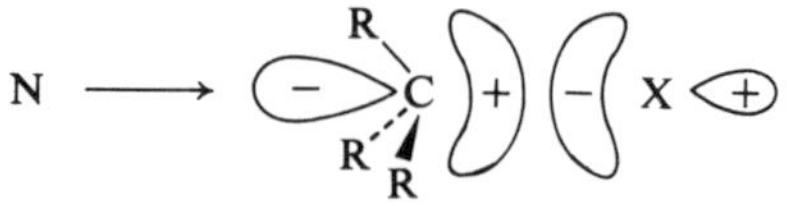

N ⟶ (−) C(R)(R)(R) (+) (−) X (+) [*3.6*]

rear of the carbon atom; approach on the front side is hindered by the repulsion between the electron pair of the nucleophile and either the bonding pairs of the C–R bonds or the lone pairs on the leaving group.

3.3.2. S_N1 reactions. The stereochemistry of some reactions which are thought to proceed by the S_N1 mechanism is shown in table 3.3. It can be seen that there is a considerable variation in the amount of racemisation observed, and that this depends both on the nature of the substrate and on the solvent employed. One explanation of this behaviour assumes that the amount of racemisation depends upon the stability, or the lifetime, of the carbonium ion (Ingold, 1969). For a relatively unstable (short-lived) carbonium ion, the reaction with the solvent occurs whilst the counter ion is close to the front side of the carbonium ion and so shields that side from attack by the solvent; product is therefore formed with predominant inversion of configuration. As the carbonium ion becomes more stable (long-lived), the counter ion is able to diffuse further away before reaction, and product is formed showing a higher degree of racemisation. An alternative description, in terms of the structural hypothesis (§1.2.3), also assumes that the stability of the carbonium ion intermediate is important in determining the stereochemistry of the product.

The results in table 3.3 indicate that there is indeed a correlation between the amount of racemisation observed and the structure of the substrate; those substrates that are expected to yield the more stable carbonium ions show the larger amounts of racemisation. Thus in the data

TABLE 3.3 *The stereochemistry of some S_N1 reactions*

Substrate	Temp. (°C)	Racemisation (%)	Net inversion[a] (%)
Acetic acid			
1-Butyl-1-d brosylate	99	–	96 ± 8
2-Octyl tosylate	75	7	93
Benzyl-1-d tosylate	25	20	80
1-Phenylethyl chloride	50	85	15
1-Phenylethyl tosylate	Room	88	12
Methanol			
2,4-Dimethyl-4-hexyl Hph[b]	Reflux	46	54
3,7-Dimethyl-3-octyl chloride	60	66	34
1-Phenylethyl chloride	70	72	28
2-Phenyl-2-butyl Hph[b]	Reflux	88	12
p-Methoxybenzhydryl Hph[b]	Warm	100	–
80% Ethanol : 20% Water			
2-Butyl brosylate	25	–	100 (ether)
Benzyl-1-d tosylate	25	{ –	93 ± 6 (ether)
		–	96 ± 5 (alcohol)
1-Phenylethyl chloride	40	{ 58.1	41.9 (ether)
		68.5	31.5 (alcohol)
80% Acetone : 20% Water			
3,7-Dimethyl-3-octyl chloride	60	79	21
1-Phenylethyl chloride	70	98	2
p-Chlorobenzhydryl chloride	25	100	–

[a] 75 per cent inversion and 25 per cent retention of configuration is equivalent to 50 per cent racemisation and 50 per cent net inversion. In the text all percentage inversions refer to net inversions.
[b] Hph = hydrogen phthalate ester.

referring to methanolysis, the amount of racemisation increases in the series, 2,4-dimethyl-4-hexyl < 1-phenylethyl < *p*-methoxybenzhydryl.

The observation of racemic products does not necessarily mean that the racemisation is a consequence of the mechanism of the substitution reaction. It might be caused by an additional process which is not connected with the substitution in any way. The acetolysis of 2-octyl tosylate (table 3.3) gives product showing 93 per cent inversion and 7 per cent racemisation. Control experiments showed, however, that this proportion of racemisation was due to racemisation of both the starting material and the product, brought about by the *p*-toluenesulphonic acid formed in the acetolysis (Streitwieser, Walsh and Wolfe, 1965). The solvolytic displacement itself proceeded with essentially complete inversion of configuration.

3.3.3. Solvolysis in mixed Solvents. The solvolysis of optically active 2,4-dimethyl-4-hexyl hydrogen phthalate in methanol [*3.7*] produces the corresponding methyl ether with about 60 per cent inversion of configuration and 40 per cent racemisation. When the methanol is diluted

$$CH_3{-}\underset{CH_2CH(CH_3)_2}{\overset{CH_2CH_3}{\underset{|}{\overset{|}{C}}}}{-}O.CO{-}C_6H_4{-}COOH \xrightarrow{MeOH} CH_3{-}\underset{CH_2CH(CH_3)_2}{\overset{CH_2CH_3}{\underset{|}{\overset{|}{C}}}}{-}OCH_3 \qquad [3.7]$$

with benzene or with chloroform the stereochemical result is unchanged, but when nitromethane is used as the co-solvent the amount of inversion depends upon the composition of the solvent (Streitwieser, 1962). For a 50 mole per cent solution of methanol and nitromethane, the methyl ether is formed with about 20 per cent inversion and 80 per cent racemisation. This result is not due to racemisation of either the starting material or the product, so that the nitromethane is therefore affecting the stereochemistry of the solvolysis reaction itself.

This behaviour can be explained in terms of a mechanism [*3.8*] in which the nitromethane is assumed to form an intermediate with the carbonium ion. Although the intermediate is readily formed it cannot lead to a stable product, and so it reacts with methanol to form the normal solvolysis product, which, however, is more racemic than the product obtained entirely by the direct route from RX. The representations MeOR and ROMe imply enantiomeric forms. An explanation in terms of

$$RX \xrightarrow{MeOH} MeOR + ROMe; \quad RX \xrightarrow{CH_3NO_2} [CH_3NO_2\cdots R^+] \xrightarrow{MeOH} MeOR + ROMe \qquad [3.8]$$

the structural hypothesis simply involves additional intermediates in which solvation of the carbonium ion by the 'inert' co-solvent makes an important contribution.

Nitromethane is not the only solvent which is capable of altering the stereochemistry of a solvolysis reaction, acetonitrile, dioxane, and acetone may all behave in a similar fashion. The common feature in all of these molecules is the presence of an unshared pair of electrons available for bonding to a carbonium ion, although the molecules are not capable of forming stable products. The data in table 3.3 for 1-phenylethyl chloride indicate that in aqueous acetone some of the hydrolysis product could be formed via an acetoxonium intermediate. In this solvent the amount of inversion is much smaller than that observed in the other solvents.

There are several examples known of solvolytic displacements in mixed solvents that occur with overall retention of configuration (Goering and Hopf, 1971). In the hydrolysis of *p*-chlorobenzhydryl

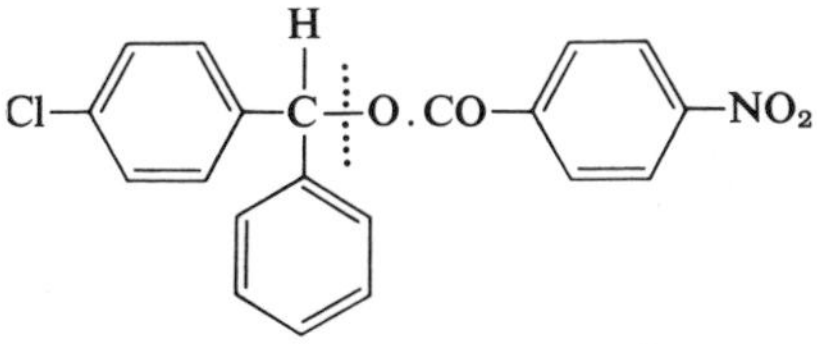

(52)

p-nitrobenzoate (52) in 90 per cent aqueous acetone, which involves exclusive alkyl–oxygen cleavage, 90 per cent of the alcohol formed is racemic but 10 per cent is formed with retention of configuration. This result can be understood in terms of the intermediate formation of an acetoxonium complex if it is assumed that both the formation and the reaction of this complex involve overall inversion of configuration. It would be a little surprising though if a benzhydryl carbonium ion were to show such a preference towards substitution with inversion. An alternative explanation, given by Goering and Hopf, is that the retention observed is the result of the preferential capture of the carbonium ion (probably at the stage of the solvent separated ion pair, cf. chapter 4) by a solvent molecule that is hydrogen bonded to the counter ion and is consequently more nucleophilic than the other molecules of solvent in the solvation shell (53).

3.3.4. Retention of configuration. It was mentioned in chapter 1 that reactions of the $S_{N}i$ type lead to retention of configuration, because the

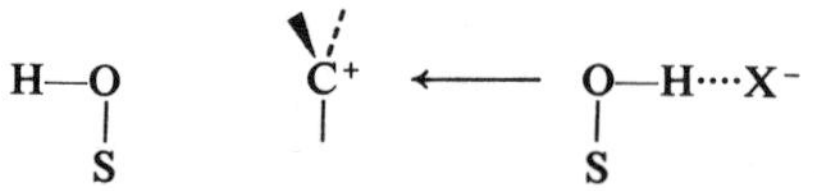

(53)

entering group is constrained by the cyclic transition state to attack the front side of the carbon atom. In many other cases retention of configuration is observed and is due to an interaction between the reacting carbon atom and a nucleophilic substituent of the substrate (neighbouring group participation, §5.1.1). This interaction, which normally may be represented as either an electrostatic interaction or a weak covalent bond, protects the rear of the carbonium ion from nucleophilic attack, so that product is formed with retention of configuration (Capon, 1964). A well-known example is the alkaline hydrolysis of optically active α-bromopropionate ion in which lactate is formed with retention of configuration [*3.9*]. In this example the carboxylate anion is suitably

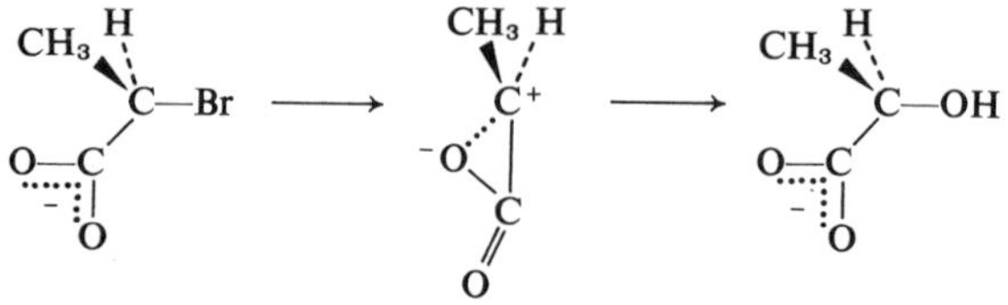

[*3.9*]

placed to interact with the reacting carbon atom, although the intermediate α-lactone has not been isolated.

The hydrolysis in 95 per cent aqueous acetone of optically active *p*-biphenylyl-1-naphthylphenylmethyl benzoate (54) produces the corresponding alcohol with 50 per cent overall retention of configuration.

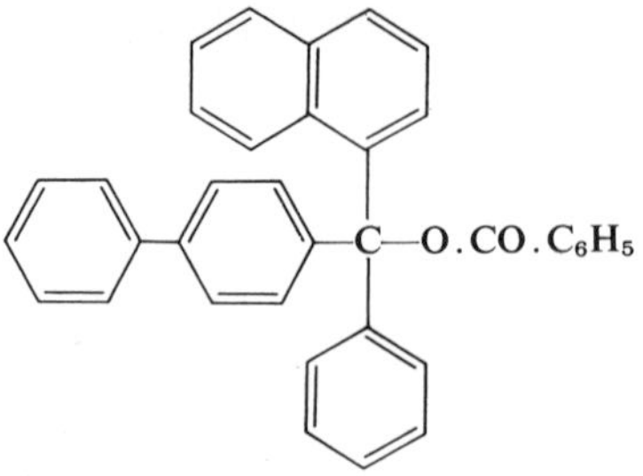

(54)

This appears, at first sight, to be another example of the type of behaviour mentioned in the preceding section. However, Murr and Santiago

(1968) have suggested that it might be due to the involvement of an asymmetric carbonium ion (55), i.e. that it is due to the stereochemical identity of the reacting carbon atom being maintained.

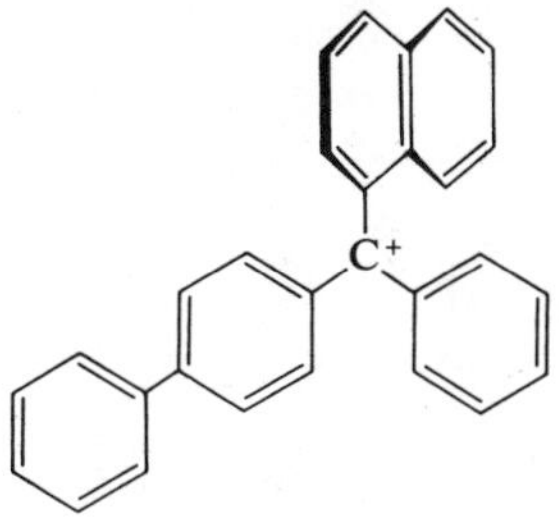

(55)

3.4. Secondary deuterium isotope effects

The isotopic substitution of deuterium for hydrogen can have an effect on the rate of a chemical reaction, whether or not the bond to hydrogen is broken during the reaction. Thus 1-phenylethyl-1-d chloride (56) undergoes solvolysis more slowly than does the corresponding protium compound.

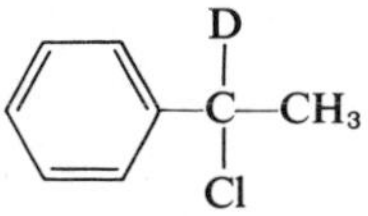

(56) 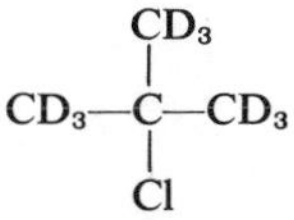 **(57)**

The ratio of solvolysis rate constants, k_H/k_D, measures the effect upon the rate of reaction caused by the substitution of deuterium for hydrogen at the α-carbon atom, and is called the secondary α-deuterium isotope effect (α-D effect). Similarly, deuterated t-butyl chloride (57) reacts more slowly than t-butyl chloride itself. In this case the isotopic substitution is at the β-carbon atoms, and the ratio of rate constants, k_H/k_D, is called the secondary β-deuterium isotope effect (β-D effect). Deuterium substitution at more remote carbon atoms may also influence the rate of reaction, but such effects are usually very small.

3.4.1. The origin of kinetic isotope effects. Using transition state theory and statistical mechanics it is possible to derive an expression (3.10), which relates the kinetic isotope effect to a ratio of complete partition

$$k_H/k_D = (Q_D/Q_H)/(Q_D^{\ddagger}/Q_H^{\ddagger}) \tag{3.10}$$

functions. In terms of this expression, the kinetic isotope effect has its origin in the effect on the molecular mass, the moments of inertia and the molecular vibrations of the substrate and of the corresponding transition state brought about by the isotopic substitution. It is possible to evaluate (3.10) (Melander, 1960), but rather than attempt a detailed analysis we shall instead develop an approximate theory sufficient for our present purposes.

The complete partition function may be expressed as a product (3.11), in which the separate terms represent the partition functions for the various degrees of freedom. For polyatomic molecules the substitution of deuterium for hydrogen will have only a minor effect on the

$$Q = q_{\text{trans}} \cdot q_{\text{rot}} \cdot q_{\text{vib}} \tag{3.11}$$

molecular masses and the moments of inertia, so that the translational and rotational partition functions will be little affected. The most important factor is then the effect of isotopic substitution on the vibrational partition function. To a good approximation the only terms which need to be considered are those which give the differences in zero point energies of the vibrational frequencies that involve the substituted hydrogens (3.12).

$$\frac{k_{\text{H}}}{k_{\text{D}}} = \frac{\exp\left[\sum_i (u_{\text{H}_i} - u_{\text{D}_i})/2\right]}{\exp\left[\sum_i (u_{\text{H}_i}{}^{\ddagger} - u_{\text{D}_i}{}^{\ddagger})/2\right]} \tag{3.12}$$

In this expression $u = h\nu/kT$, where ν is the fundamental frequency in wavenumbers (cm^{-1}), T is the absolute temperature, h is Planck's constant, and k is Boltzmann's constant. The summations are over those frequencies associated with the substituted atoms.

Equation (3.12) may be simplified to give (3.13) by making use of the empirical relation $\nu_{\text{D}} = \nu_{\text{H}}/1.35$ (Streitwieser, Jagow, Fahey and Suzuki, 1958). We see from (3.13) that all normal secondary isotope

$$k_{\text{H}}/k_{\text{D}} = \exp\left[(0.13h/kT)\sum_i (\nu_{\text{H}_i} - \nu_{\text{H}_i}{}^{\ddagger})\right] \tag{3.13}$$

effects (i.e. $k_{\text{H}}/k_{\text{D}} > 1$) arise when the relevant hydrogen vibrational frequencies are smaller in the transition state than in the ground state, i.e. when formation of the transition state leads to a weakening of the bond to the hydrogen atom. A decrease in frequency of 300 cm^{-1}, for example, produces an isotope effect of $k_{\text{H}}/k_{\text{D}} \approx 1.15$. In terms of this approximate theory, therefore, the origin of the isotope effect is the differential change in the zero point energies of the C–H and C–D vibrations accompanying formation of the transition state. The transition state is at a saddle point on a potential energy surface and one vibrational degree of freedom is unstable to motion in the direction of the reaction co-ordinate; for an S_N1 process this would be the vibration which causes the heterolysis of the C–X bond. All the other vibrations may be treated as though the transition state were a normal molecule. This is represented in fig. 3.1, in which the potential energy curves for

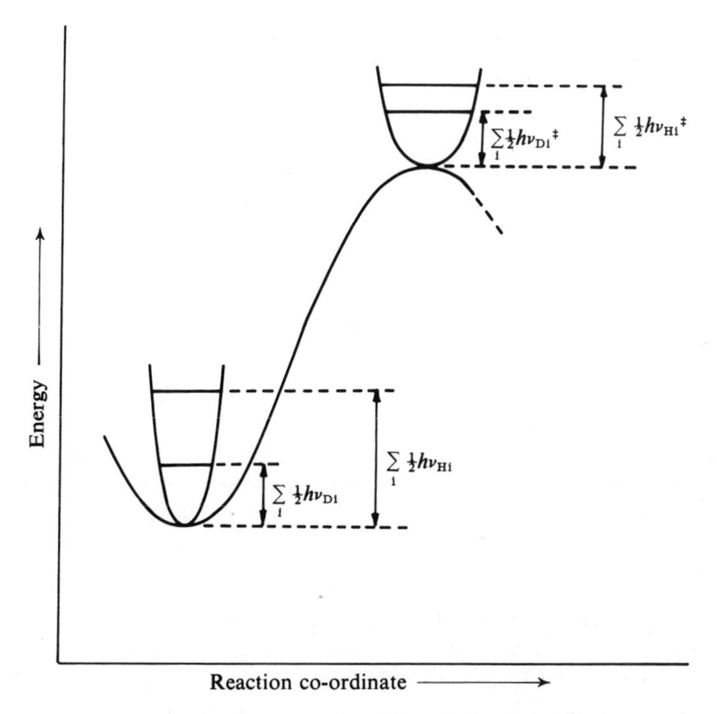

Fig. 3.1 A representation of the zero point energies responsible for a secondary deuterium isotope effect for an S_N1 process.

the hydrogen vibrations lie in planes which are at right angles to the plane containing the reaction co-ordinate. The zero point energy levels of the hydrogen vibrations responsible for the secondary deuterium isotope effect are indicated.

The smaller vibrational frequencies in the transition state imply smaller zero point energies for the vibrations, compared with those of the ground state. Since a C–D bond has a lower zero point energy than a C–H bond, less zero point energy is lost at the transition state for the deuterium compound than for the corresponding hydrogen compound. The total energy difference between the ground state and the transition state is therefore greater for the deuterium compound which should thus react more slowly.

3.4.2. The α-D isotope effect. The use of secondary deuterium isotope effects in mechanistic studies arises from the fact that the magnitude of the effect has been found to be related to the mechanisms of nucleophilic substitutions. Values of k_H/k_D close to unity are found for most S_N2 reactions, and in the range 1.10 to 1.23 for many S_N1 reactions. These differences may be rationalised in terms of the vibrational characteristics of the ground state (58) and those of the different types of transition states (59) and (60) (Streitwieser, Jagow, Fahey and Suzuki, 1958). The vibration of importance appears to be the HCX bending

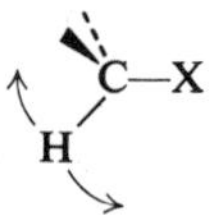

C—X, H **(58)** C + X^-, H **(59)** N····C····X, H **(60)**

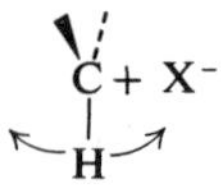

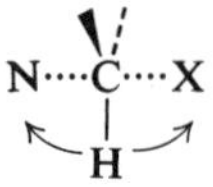

mode in the ground state, which becomes a sort of out-of-plane bending in the transition state. For the S_N1 transition state (59) the breaking of the C–X bond and the change of hybridisation at the reacting carbon atom lead to an easier bending motion for hydrogen than is found in the ground state. For the S_N2 transition state (60) the partial bonding of both N and X to the reacting carbon atom impedes the bending of the hydrogen, and there may be no overall change in the ease of vibration compared with that in the ground state.

The data in table 3.4 show that for a variety of substrates, leaving groups and nucleophiles, the value of k_H/k_D for an S_N2 process is usually within 5 per cent of unity. The spread of values about unity can be attributed to the slightly different bonding in the transition states. A small normal isotope effect should be observed with a loose transition state, in which the hydrogen bending motion is slightly easier than that in the ground state. Conversely, a small inverse isotope effect can be expected if the reaction involves a tight transition state (§2.1.2). The magnitude of the α-D effect can therefore be expected to depend upon the relative nucleophilicities of the groups X and N, and such a trend is evident in the data in table 3.4 (Seltzer and Zavitsas, 1967).

TABLE 3.4 *Secondary α-deuterium isotope effects for some* S_N2 *reactions*[a]

Substrate	Nucleophile	Solvent	Temp. (°C)	k_H/k_D
CD_3I	H_2O	H_2O	70	0.87
CD_3Br	H_2O	H_2O	80	0.90
CD_3Cl	H_2O	H_2O	90	0.92
CD_3OTs	H_2O	H_2O	70	0.96
CD_3OBs	MeOH	MeOH	70	0.94
CD_3I	Et_3N	C_6H_6	50	0.88
CD_3I	C_5H_5N	C_6H_6	50	0.92
CD_3I	I^-	MeOH	20	1.05
CD_3Br	$S_2O_3^{2-}$	$EtOH:H_2O$	25	1.03
CH_3CD_2Br	$S_2O_3^{2-}$	$EtOH:H_2O$	25	1.08
CH_3CD_2OTs	H_2O	H_2O	54	1.04
$C_2H_5CD_2I$	H_2O	H_2O	90	1.01
$(CH_3)_2CDBr$	H_2O	H_2O	60	1.07
$(CH_3)_2CDBr$	EtO^-	EtOH	25	1.00

[a] Data from Seltzer and Zavitsas (1967).

The data referring to S_N1 solvolytic displacements (table 3.5) illustrate the important features of α-D effects for limiting reactions; they are not very dependent upon the structure of the substrate or upon the solvent, but they show a characteristic variation with the nature of the leaving group. Arenesulphonates give the largest values of k_H/k_D (1.23), followed by chlorides (1.15), bromides (1.12), and iodides (1.09) (Shiner and Dowd, 1971). This leaving group effect can be quantitatively accounted for in terms of the differences in the HCX bending vibrations of the ground states.

3.4.3. The β-D isotope effect. The substitution of deuterium for β-hydrogen leads to a decrease in the rates of many S_N1 reactions. In terms of the theory developed above, this effect is explained by assuming a decrease in the frequencies of those vibrations involving the β-hydrogens on passage to the transition state. Of the factors, inductive, steric, and hyperconjugative, that have been considered as being responsible for these changes, it appears that the last mentioned effect is the most important (Shiner, 1970).

Two observations support this conclusion. First, the β-D effect shows

TABLE 3.5 *Secondary α-deuterium isotope effects for some S_N1 solvolytic displacements at 25 °C*

Substrate	Solvent	k_H/k_D [a]
$(CH_3)_2CDOTs$	Trifluoroacetic acid	1.22
p-Me.C_6H_4.CD_2Cl	97% TFE [b]	1.142
p-PhO.C_6H_4.$CDClCH_3$	93% acetone [c]	1.147
p-PhO.C_6H_4.$CDBrCH_3$	93% acetone	1.126
C_6H_5.$CDClCH_3$	80% ethanol [d]	1.146
C_6H_5.$CDBrCH_3$	80% ethanol	1.122
C_6H_5.$CDClCH_3$	50% ethanol	1.153
$CH_3C{\equiv}C$.$CDOTsCH_3$	60% ethanol	1.213
$CH_3C{\equiv}C$.$CDOTsCH_3$	70% TFE	1·226
$CH_3C{\equiv}C$.$CDBrCH_3$	50% ethanol	1.101
$CH_3C{\equiv}C$.$CDBrCH_3$	70% TFE	1.123
$CH_3C{\equiv}C$.$CDICH_3$	50% ethanol	1.087
$CH_3C{\equiv}C$.$CDICH_3$	70% TFE	1.089
D OTres [e]	50% Ethanol	1.225
	70% TFE	1.225
	97% TFE	1.228

[a] k_H/k_D per α-D.
[b] 97 w/w per cent 2,2,2-trifluoroethanol:3 w/w per cent water, etc.
[c] 93 v/v per cent acetone:7 v/v per cent water.
[d] 80 v/v per cent ethanol:20 v/v per cent water, etc.
[e] −OTres = 2,2,2-trifluoroethylsulphonate.
Data from Shiner and Dowd (1971); Shiner and Fisher (1971); Shiner (1970).

a conformational dependence; the magnitude of the isotope effect depends upon the dihedral angle, θ, between the β-C–H bond and the vacant p-orbital on the reacting carbon atom (61). Thus isotope effects

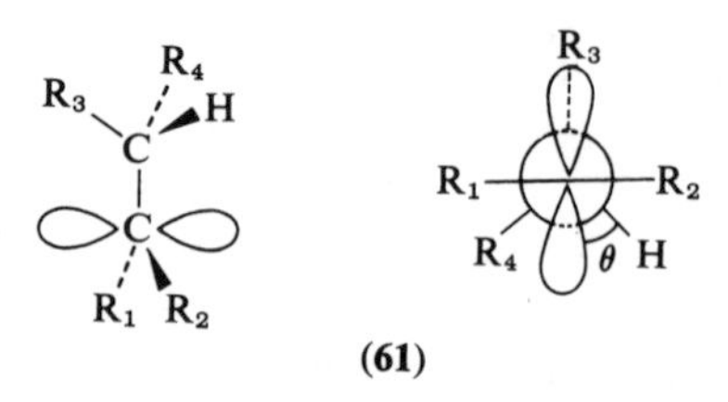

(61)

have been measured for the solvolysis of the bicyclic tertiary chlorides (62) and (63) in 60 per cent aqueous ethanol. In (62) the dihedral angle is constrained to be 90° and so hyperconjugation is stereoelectronically prevented, $k_H/k_D = 0.986$. In (63) the dihedral angles should be close to 30°, and a small normal β-D effect is observed, $k_H/k_D = 1.14$.

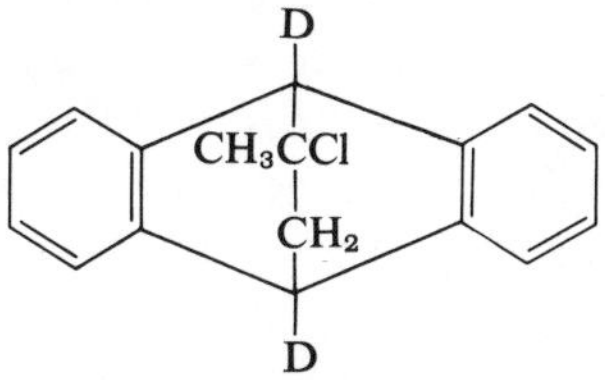

(62) 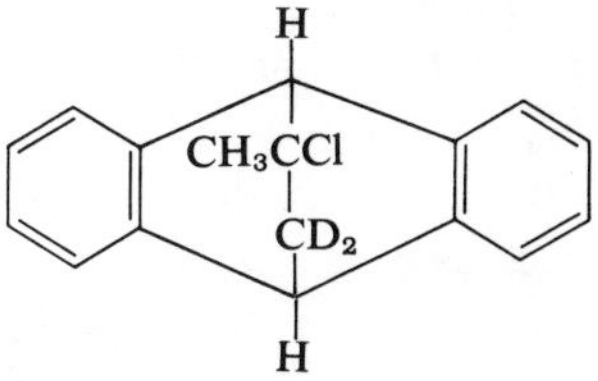 (63)

Secondly, the β-D effect is observed to decrease in a series of substrates as the groups R_1 and R_2 (61) become more effective at stabilising the positive charge on the α-carbon atom, i.e. as the need for hyperconjugative stabilisation becomes less important (table 3.6). In the 1-phenylethyl series and the tertiary propargyl series the magnitude of the β-D effect decreases with increasing stability of the carbonium ion.

The maximum β-D isotope effect has been estimated to be *ca* 1.46 per —CD_3 (Shiner, 1970) and for the S_N1 mechanism any value between this limit and unity may be observed. The β-D effect for an S_N2 reaction is also close to unity, so that the magnitude of the β-D effect, by itself, is a poor indicator of the mechanism. However, when used with the α-D

TABLE 3.6 *Secondary β-deuterium isotope effects for some* S_N1 *solvolytic displacements at 25 °C*

Reactant	Solvent	k_H/k_D [a]
p-PhO.C_6H_4.$CHClCD_3$	95% ethanol [b]	1.164
p-Me.C_6H_4.$CHClCD_3$	95% ethanol	1.200
p-Me.C_6H_4.$CHClCD_3$	80% ethanol	1.199
$C_6H_5CHClCD_3$	80% ethanol	1.223
C_6H_5.$CHClCD_3$	50% ethanol	1.224
C_6H_5.$CHBrCD_3$	80% ethanol	1.220
$(CH_3)_2$.$CClCD_3$	60% ethanol	1.220
$HC{\equiv}C.CBr(CD_3)_2$	80% ethanol	1.357
$CH_3C{\equiv}C.CCl(CD_3)_2$	80% ethanol	1.286
$^-{:}C{\equiv}C.CBr(CD_3)_2$	80% ethanol	1.144
$CH_3C{\equiv}C.CHOTsCD_3$	70% TFE [c]	1.281
$CH_3C{\equiv}C.CHBrCD_3$	70% TFE	1.280
$CH_3C{\equiv}C.CHICD_3$	70% TFE	1.283

[a] k_H/k_D per β-CD_3 [$= (k_H/k_{\beta\text{-}D_6})^{1/2}$].
[b] 95 v/v per cent ethanol:5 v/v per cent water.
[c] 70 w/w per cent trifluoroethanol:30 w/w per cent water.
Data from Shiner (1970); Shiner and Dowd (1971).

effect it provides useful information about the structure of the transition state. In the 1-phenylethyl series (table 3.6) the β-D effect varies with the structure of the alkyl group, although the corresponding α-D effect is almost constant. The conclusions drawn are that all the substrates react by the same limiting S_N1 mechanism, and that the different β-D effects reflect differences in the carbonium-ion stabilities. In the secondary propargyl series the reverse situation is encountered; the β-D effects remain constant, within experimental error, while the α-D effects vary. The variation in the α-D effect is explained in terms of a leaving-group effect.

3.4.4. Remote deuterium isotope effects. The substitution of deuterium for hydrogen at positions more remote than the β-carbon atom has only a minor effect upon the rate of an S_N reaction, unless a mechanism exists whereby the remote carbon atom can interact strongly with the reaction centre. The tertiary chloride (64), for example, undergoes solvolysis in 80 per cent ethanol at 25 °C and gives an isotope effect $k_H/k_D = 0.983$.

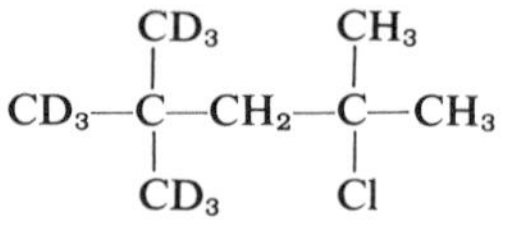

CD₃ CH₃
CD_3—C—CH_2—C—CH_3
CD₃ Cl

(64)

When the remote site is in an unsaturated system at a position capable of a hyperconjugative interaction with the reaction centre, an appreciable isotope effect may result. Thus the substitution of deuterium for hydrogen in the *p*-methyl group of (65) leads to a value of $k_H/k_D = 1.10$ for acetolysis at 50 °C.

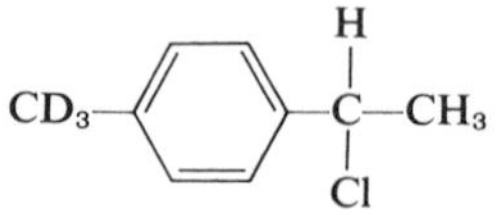

H
CD_3—C₆H₄—C—CH_3
Cl

(65)

A large γ-D effect has also been observed in the acetolysis of *exo*-2-norbornyl brosylate (66), $k_H/k_D = 1.097$ (Jerkunica, Borčić and Sunko, 1967). The corresponding *endo*-compound (67) showed no appreciable effect, $k_H/k_D = 0.998$.

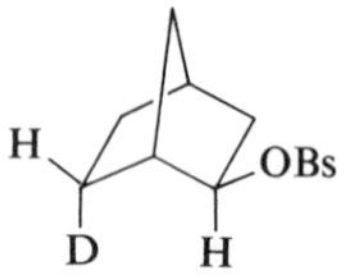

H OBs
D H

(66)

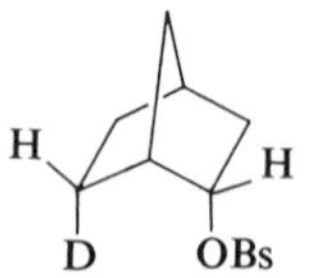

H H
D OBs

(67)

Similar results were obtained when the deuterium was in the 6-*exo*-position. These results indicate that an interaction occurs between the 6-position and the reacting centre in the solvolysis of (66) but not of (67). This is in agreement with the idea that the rate-limiting step in the solvolysis of (66) involves the non-classical, bridged, norbornyl cation (68) (§5.3).

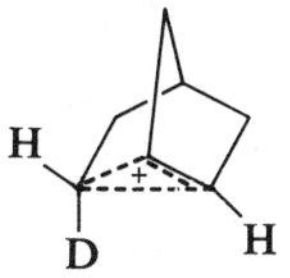

(68)

3.5. Activation parameters

The temperature dependence of the rates of most reactions can be described by the Arrhenius equation (3.14), in which E is the Arrhenius

$$\mathrm{d}(\ln k)/\mathrm{d}T = E/RT^2 \tag{3.14}$$

activation energy. Integration of (3.14) gives (3.15), in which the constant of integration A is called the pre-exponential term or frequency

$$\ln k = -E/RT + \ln A \tag{3.15}$$

factor. Both E and A are assumed to be independent of the temperature and may be evaluated from kinetic measurements, either graphically from a plot of log k against $1/T$, or, for a small temperature range, from (3.16) and (3.17).

$$E = (T_1T_2/(T_1 - T_2))R \ln(k_1/k_2) \tag{3.16}$$

$$\ln A = (T_1 \ln k_1 - T_2 \ln k_2)/(T_1 - T_2) \tag{3.17}$$

Within the transition state formulation the rate constant for a reaction is related to the Gibbs free energy of activation, $\Delta G^{\ddagger}$, by (3.18). In this

$$k = (\mathbf{k}T/h) \exp(-\Delta G^{\ddagger}/RT) \tag{3.18}$$

equation, and those which follow, the pseudo-thermodynamic functions all refer to a suitably chosen standard state. It is also assumed that solutions are dilute, so that all activity coefficients are unity. Using the relation (3.19), we can obtain (3.20), from which can be obtained the result (3.21). At normal reaction temperatures, therefore, the enthalpy of activation, $\Delta H^{\ddagger}$, is almost equal to the Arrhenius energy of activation.

$$\Delta G^{\ddagger} = \Delta H^{\ddagger} - T\Delta S^{\ddagger} \tag{3.19}$$

$$k = (kT/h)\exp(-\Delta H^{\ddagger}/RT)\exp(\Delta S^{\ddagger}/R) \tag{3.20}$$

$$\Delta H^{\ddagger} = E - RT \tag{3.21}$$

It can also be shown that the entropy of activation, $\Delta S^{\ddagger}$, is related to $\ln A$ by the expression (3.22). The terms $\Delta H^{\ddagger}$ and $\Delta S^{\ddagger}$ are both depen-

$$\Delta S^{\ddagger} = R\ln A - R(\ln(kT/h) + 1) \tag{3.22}$$

dent upon the temperature and it is therefore important when considering differences in activation parameters, particularly when these differences are small, to compare data referring to the same temperature.

3.5.1. The enthalpy and entropy of activation. The interpretation of the magnitudes of activation parameters for reactions in solution is complicated by the fact that the change in the rate of reaction with the temperature is determined partly by the mechanism of the reaction and partly by the nature of the solvent. A limited selection of the available data (table 3.7) is sufficient to illustrate the main points.

It is immediately apparent that the value of $\Delta H^{\ddagger}$, by itself, gives little information about the mechanism, although there are indications that for some non-solvolytic S_N2 reactions $\Delta H^{\ddagger}$ is smaller than for other nucleophilic substitutions. The value of $\Delta S^{\ddagger}$, on the other hand, seems to show a systematic variation with mechanism and two trends are discernible. First, in most solvents the values of $\Delta S^{\ddagger}$ are negative, with S_N2 reactions generally having numerically larger values than S_N1 reactions. Second, for S_N1 reactions in water, or mixed aqueous organic solvents containing mostly water, the values of $\Delta S^{\ddagger}$ are positive. The first four entries in table 3.7, in particular, illustrate the large variation in the value of $\Delta S^{\ddagger}$ that is often observed. For the case in point, the solvolysis of t-butyl chloride, the mechanism of the reaction is thought to remain the same in each solvent and the values of $\Delta H^{\ddagger}$ are fairly similar in all cases. The value of $\Delta S^{\ddagger}$ is dependent upon the nature of the solvent and thus appears to be a measure of the structural changes that occur in the solvent as well as in the reactants. The more negative values of $\Delta S^{\ddagger}$ for S_N2 reactions than for S_N1 reactions are consistent

TABLE 3.7 *Enthalpies and entropies of activation for some S_N reactions*

Reaction		Solvent	Mechanism	Temp. (°C)	$\Delta H^{\ddagger}$ (kJ mol^{-1})	$\Delta S^{\ddagger}$ (J K^{-1} mol^{-1})
$(CH_3)_3CCl$	solvolysis[a]	80% EtOH	S_N1	25	93.47	−27.6
$(CH_3)_3CCl$	solvolysis[a]	30% EtOH	S_N1	25	86.06	−0.42
$(CH_3)_3CCl$	solvolysis[a]	80% dioxane	S_N1	25	92.17	−48.1
$(CH_3)_3CCl$	solvolysis[a]	Water	S_N1	25	97.15	+51.0
t-Pentyl Cl	solvolysis[b]	Water	S_N1	10	91.96	+42.7
$C_6H_5CH_2CH_2OBs$	solvolysis[c]	AcOH	S_N1	75	104.18	−63.6
2,4-$(CH_3O)_2C_6H_3CH_2CH_2OBs$	solvolysis[c]	AcOH	S_N1	75	94.14	−37.7
CH_3Cl	solvolysis[b]	Water	S_N2	75	100.42	−51.5
CH_3OBs	solvolysis[b]	Water	S_N2	40	87.57	−45.6
$C_6H_5CH_2Cl$	solvolysis[b]	Water	S_N2	40	87.11	−46.0
p-$NO_2C_6H_4CH_2Cl$ + pyridine[d]		MeOH	S_N2	25	100.8	−18.6
p-$NO_2C_6H_4CH_2Cl$ + pyridine[d]		DMF	S_N2	25	59.8	−107.1
$C_6H_5CH_2Cl$ + pyridine[d]		MeOH	S_N2	25	78.7	−84.1
$C_6H_5CH_2Cl$ + pyridine[d]		DMF	S_N2	25	56.5	−157.7
p-$MeC_6H_5CH_2Cl$ + pyridine[d]		MeOH	S_N2	25	69.9	−105.6
p-$MeC_6H_5CH_2Cl$ + pyridine[d]		DMF	S_N2	25	54.4	−157.7

[a] Data from Winstein and Fainberg (1957). [b] Data from Robertson (1967). [c] Data from Winstein and Heck (1956).
[d] Data from Haberfield, Nudelman, Bloom, Romm and Ginsberg (1971).

with this idea, since S_N1 reactions can be expected to disrupt the solvent structure to a larger extent than S_N2 reactions (a disruption of solvent structure, i.e. a decrease in the solvent–solvent interactions, will make a positive contribution to $\Delta S^\ddagger$).

The last six entries in table 3.7 show the variations in $\Delta S^\ddagger$ and in $\Delta H^\ddagger$ for three Menschutkin reactions on transferring from methanol to dimethylformamide (DMF). The change is accompanied by a large decrease in the values of $\Delta H^\ddagger$, yet the rates of reaction are hardly affected. This is because an equally large decrease in the value of $\Delta S^\ddagger$ compensates for the change in $\Delta H^\ddagger$, cf. (3.20). This type of behaviour is quite common and in many instances it has the precision of a linear relationship (3.23). In this isokinetic relationship the parameter $\beta(\geq 0)$ has the

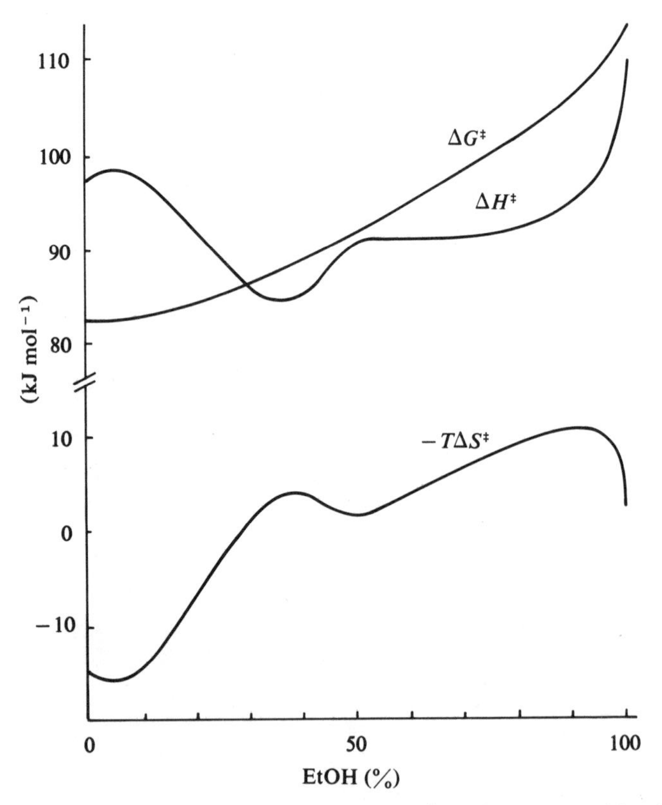

Fig. 3.2 The variation of activation parameters with solvent composition for the solvolysis of t-butyl chloride in ethanol–water solvents at 25 °C. Data from Winstein and Fainberg (1957).

dimensions of absolute temperature and it is the point at which $\delta\Delta G^{\ddagger} = 0$ (Leffler and Grunwald, 1963).

$$\delta\Delta H^{\ddagger} = \beta\delta\Delta S^{\ddagger} \tag{3.23}$$

For some reaction series the values of $\Delta S^{\ddagger}$ remain fairly constant and variations in $\Delta G^{\ddagger}$ are largely the result of changes in $\Delta H^{\ddagger}$ (isoentropic series). Other series are known in which the reverse situation obtains, and changes in $\Delta G^{\ddagger}$ are mainly due to changes in $\Delta S^{\ddagger}$, while $\Delta H^{\ddagger}$ remains fairly constant (isoenthalpic series). For reactions in mixed solvents the behaviour of $\Delta S^{\ddagger}$ and $\Delta H^{\ddagger}$ can become quite complicated, particularly when water is one of the components. One of the best-known examples is the solvolysis of t-butyl chloride in ethanol–water solvents (Winstein and Fainberg, 1957), in which, although $\Delta G^{\ddagger}$ shows a fairly steady increase with decreasing water content, both $\Delta H^{\ddagger}$ and $\Delta S^{\ddagger}$ exhibit remarkable variations (fig. 3.2). Equally remarkable is the graph obtained when the values of $\Delta H^{\ddagger}$ are plotted against those of $\Delta S^{\ddagger}$ (fig. 3.3). This apparently random behaviour can be understood (see p. 87) in terms of a variation in the structure of the solvent.

3.5.2. Analysis of activation parameters. The Menschutkin reaction [*3.10*] is much faster in dipolar aprotic solvents than in non-polar solvents. A change from the former to the latter type of solvent is therefore

$$O_2N\text{—}C_6H_4\text{—}CH_2Cl + NMe_3 \longrightarrow O_2N\text{—}C_6H_4\text{—}CH_2\overset{+}{N}Me_3 + Cl^- \qquad [3.10]$$

accompanied by an increase in the Gibbs free energy of activation, $\delta\,\Delta G^{\ddagger}$. This increase may be calculated from the rate constants for the reaction in the two solvents. It is of greater interest, however, to determine whether this change is due to a ground state effect or to a transition state effect (3.24). The term $\delta\Delta G^{t}$ represents the change in the Gibbs free

$$\delta\Delta G^{\ddagger} = \delta\Delta G^{t} - \delta\Delta G^{g} \tag{3.24}$$

energy of the transition state on transferring it from one solvent to the other, and $\delta\Delta G^{g}$ is the corresponding term for the ground state. The term $\delta\Delta G^{g}$ may be calculated for the present example from the known Henry's law constants for trimethylamine and the molar solubilities of *p*-nitrobenzyl chloride. The term $\delta\Delta G^{t}$ may therefore be evaluated from (3.24). For reaction [*3.10*] it is found that the effect of solvent on the rate (i.e. on $\Delta G^{\ddagger}$) is largely a transition state effect (Abraham, 1969).

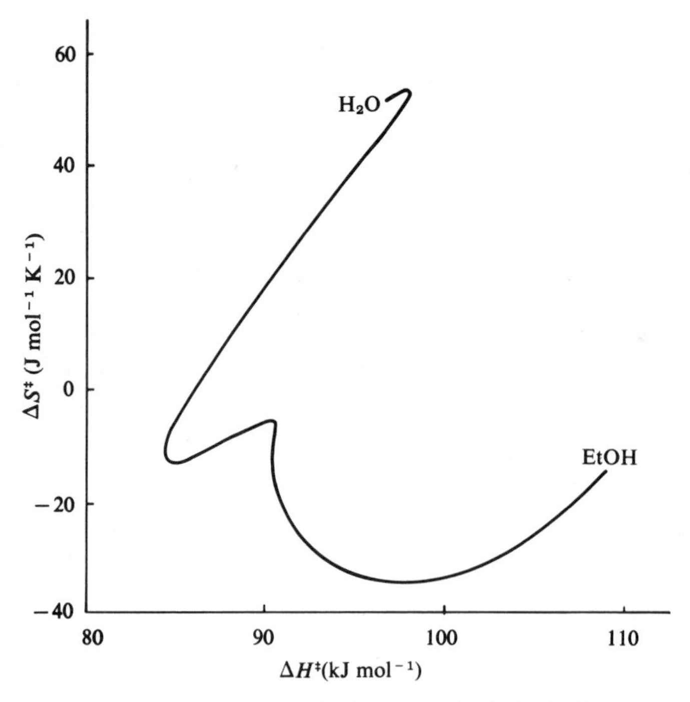

Fig. 3.3 The variation of $\Delta S^{\ddagger}$ with $\Delta H^{\ddagger}$ for the solvolysis of t-butyl chloride in ethanol–water solvents at 25 °C. Data from Winstein and Fainberg (1957).

A more detailed analysis is possible if the enthalpy of transfer of the reactants, $\delta\Delta H^{g}$, is known. In this case the enthalpy of transfer of the transition state, $\delta\Delta H^{t}$, may be calculated from (3.25), in which $\delta\Delta H^{\ddagger}$ is the difference in the enthalpies of activation for the reaction in the two solvents of interest. Values of $\delta\Delta H^{g}$ may be obtained from calorimetric

$$\delta\Delta H^{\ddagger} = \delta\Delta H^{t} - \delta\Delta H^{g} \tag{3.25}$$

measurements of the enthalpies of solution of the reactants (Arnett, Bentrude, Burke and Duggleby, 1965). Such a complete analysis of the activation parameters has been achieved for the solvolysis of t-butyl chloride in aqueous ethanol, over the range of solvent composition from ethanol to 40 per cent aqueous ethanol (Arnett, Bentrude and Duggleby, 1965). The results clearly show that the complex behaviour of $\Delta H^{\ddagger}$ and $\Delta S^{\ddagger}$ (cf. fig. 3.3) is associated with solvent effects on the ground state, while the transition state shows a much simpler type of behaviour; a plot of $\delta\Delta S^{t}$ against $\delta\Delta H^{t}$ is approximately linear.

3.5.3. Activation parameters and solvent structure. The value of $\Delta S^{\ddagger}$ is determined, in part, by the changes in the solvation of the reactants and the transition state and, in part, by the change in the solvent structure that accompanies the activation process. The same factors also play an important part in determining the magnitude of the heat capacity of activation, $\Delta C_p^{\ddagger}$ ($\equiv d(\Delta H^{\ddagger})/dT$). The majority of the measurements of $\Delta C_p^{\ddagger}$ refer to the solvolyses of initially neutral substrates in water (Robertson, 1967). Values in the range -75 to -200 J K^{-1} mol^{-1} are usually observed for those reactions occurring by the S_N2 mechanism and in the range -200 to -400 J K^{-1} mol^{-1} for the S_N1 mechanism. It is assumed that these values of $\Delta C_p^{\ddagger}$ are determined by the considerable reorganisation of the solvent that results from the development of charge in the transition state. Support for this idea comes from the observation that the value of $\Delta C_p^{\ddagger}$ is small when the leaving group is initially positively charged; the hydrolysis of the t-butyl dimethylsulphonium ion in water has the particularly small value of -33 J K^{-1} mol^{-1} for $\Delta C_p^{\ddagger}$.

The negative values of $\Delta C_p^{\ddagger}$ require that $\Delta S^{\ddagger}$ and $\Delta H^{\ddagger}$ both decrease as the temperature increases, and two explanations have been suggested to account for this behaviour. For solvolysis in water, or in aqueous organic solvents containing mostly water, it is assumed that the factor responsible is the disruption of the ground state solvent structure. For reactions in aqueous organic solvents, on the other hand, the increased solvation of the transition state is assumed to be the more important factor. It should be pointed out, however, that the suggestion has also been made that the observed variation of $\Delta H^{\ddagger}$ with the temperature, for solvolysis reactions, might be due to the effect of a change of temperature on the partitioning of an ion-pair intermediate, cf. chapter 4. It has been shown that such an interpretation might apply when the rate constants of the various pathways for the destruction of the ion pair are of comparable magnitude (Albery and Robinson, 1969).

The behaviour in water seems to be a result of the particularly strong solvent–solvent interactions which confer great stability on the water structure. It will be recalled (table 3.7) that for S_N1 reactions in water the value of $\Delta S^{\ddagger}$ is positive, a result which is also consistent with the idea of a loss of solvent structure on formation of the transition state. A variation in the water structure affords an explanation of the complex behaviour of $\Delta H^{\ddagger}$ and $\Delta S^{\ddagger}$ (figs 3.2 and 3.3) observed for the solvolysis of t-butyl chloride in solutions of aqueous ethanol at high concentrations of water (Arnett, Bentrude, Burke and Duggleby, 1965).

3.6. Linear free energy relationships

3.6.1. The Hammett equation. The Hammett equation (3.26), has been used to correlate data for a large number of well characterised reaction

$$\log (k/k_0) = \sigma\rho \tag{3.26}$$

series. The significance of the signs and magnitudes of the parameters involved are fairly well understood, and mechanistic interpretations are now often based on the values of the parameters determined for a new reaction series. The information of interest is the type of σ-value needed to produce a linear correlation with the experimental results and the magnitude of the ρ-value (Johnson, 1973).

The σ-parameter in (3.26) is the substituent constant, which depends only upon the substituent and not on the particular reaction being considered. As defined, it measures the effect of the substituent on the ionisation of benzoic acid in aqueous solution, $\sigma = (\log K_a - \log K_a{}^0)$. From this definition of σ, electron withdrawing substituents (i.e. acid strengthening) will have positive values, and electron donating substituents will have negative values. In addition to the σ-values there are now available several other sets of substituent constants, e.g. σ^0, σ^n, σ^+ (cf. Wells, 1968), all of which are independent of the reaction series, the temperature, and reasonably independent of the solvent.

The reaction constant, ρ, is simply a proportionality constant, but by defining its value as unity for the standard reaction, it provides a measure of the sensitivity of a given reaction to substituent effects, relative to that of the dissociation of benzoic acids. It follows from the definition of σ that reactions accelerated by electron withdrawing substituents will have positive ρ-values, and that those slowed down by electron withdrawing substituents will have negative ρ-values. The magnitude of ρ gives information about the amount of charge developed at the reaction centre during reaction; a large ρ-value implies a large change. ρ-Values are dependent on both the solvent and the temperature. The latter dependence can have important consequences in the assignment of reaction mechanisms, because the ρ-value may change sign with a change of temperature.

The application of the Hammett equation to nucleophilic substitution reactions is limited to those cases in which a benzene ring is attached directly to the reaction centre. For S_N1 reactions the experimental results are well correlated with the σ^+-values of the substituents in the benzene ring and the reaction constants have large negative values. This is not too surprising bearing in mind the defining reaction for the σ^+ scale, the solvolysis of t-cumyl chlorides (69) in 90 per cent aqueous acetone, a reaction for which $\rho = -4.54$. With some substrates the kinetic results

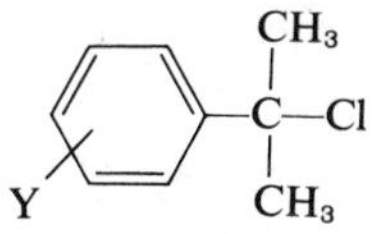

(69)

are found not to fit a single linear correlation; a plot of the logarithms of the relative rates against σ^+ then gives either a smooth curve or two separate linear correlations. The latter type of behaviour is illustrated by the results for the acetolysis of substituted benzyl tosylates at 40 °C (Streitwieser, Hammond, Jagow, Williams, Jesaitis, Chang and Wolf, 1970) (fig. 3.4). The correlation for electron donating substituents is typical of those observed for S_N1 reactions, the value of $\rho = -5.71$ indicates that these substrates react by a single mechanism which has a high degree of carbonium-ion character. The results for electron withdrawing substituents, however, which correlate with a ρ-value of -2.33, indicate that with these substituents a mechanism requiring less carbonium-ion character is involved.

It was seen in chapter 1 that the isotopic exchange between substituted benzhydryl thiocyanates and ionic thiocyanate, under certain conditions, takes place by concurrent S_N1 and S_N2 mechanisms. The first-order rate constants (table 1.2) show a linear correlation with σ^+-values giving a ρ-value of -4.5. On the other hand, the second-order rate constants are not well correlated, and, in fact, show that the S_N2 reaction is not particularly sensitive to ring substituents (fig. 3.5). This last result implies that $\rho \approx 0$, which could arise if the transition state for the S_N2 reaction is symmetrical with the excess charge being shared equally by the incoming and leaving groups (70). In such a situation there is little change

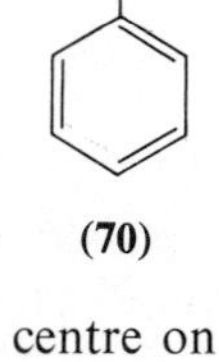

(70)

in the charge at the reaction centre on passage to the transition state. The points for the S_N2 reactions (fig. 3.5) actually lie on a shallow

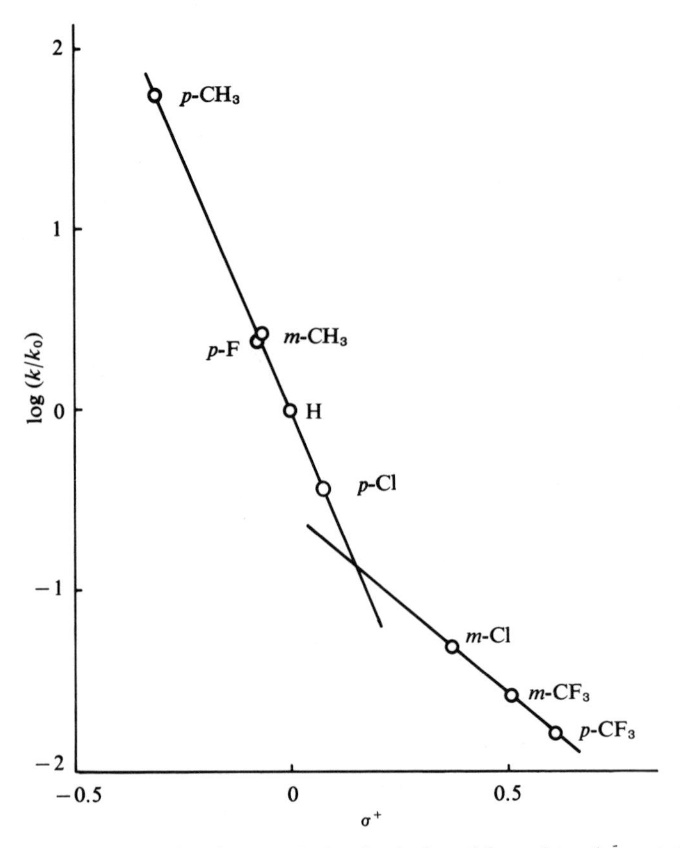

Fig. 3.4 $\rho\sigma^+$ correlation for acetolysis of substituted benzyl tosylates at 40 °C. Data from Streitwieser, Hammond, Jagow, Williams, Jesaitis, Chang and Wolf (1970).

U-shaped curve, which may indicate that the nature of the transition state varies with the substituents in the phenyl rings (§1.3). In general, the value of ρ for an S_N2 reaction depends on whether a loose (small negative ρ) or a tight (small positive ρ) transition state is involved.

3.6.2. The Taft equation. The rates of many reactions involving aliphatic systems are found to be correlated by (3.27), which bears a formal similarity to the Hammett equation. The polar substituent constants,

$$\log (k/k_0) = \sigma^* \rho^* \tag{3.27}$$

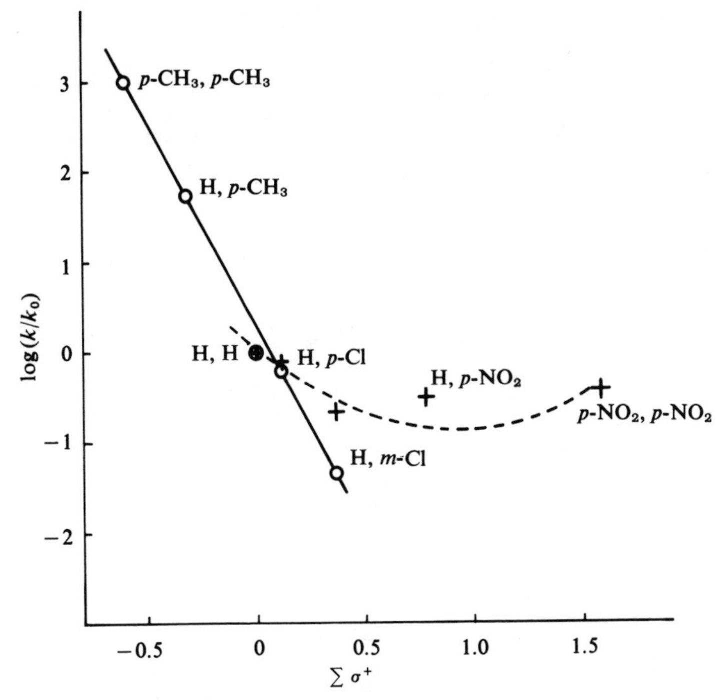

Fig. 3.5 $\rho\sigma^+$ correlation for isotopic exchange of substituted benzhydryl thiocyanates in acetonitrile at 70 °C. ○ first-order, + second-order rate constants. Data from Ceccon, Papa and Fava (1966).

σ^*, were defined by Taft (3.28), as the difference in the logarithms of the

$$\sigma^* = (1/1.25)[\log (k/k_0)_B - \log (k/k_0)_A] \tag{3.28}$$

relative rates for the base- and acid-catalysed hydrolyses of an ester in which the substituent was attached to the carbonyl group; the reference compound was chosen to be the corresponding methyl compound. The factor (1/1.25) was introduced to put the σ^*-values on a similar scale to the Hammett σ-values. This treatment represents an attempt to derive a substituent constant dependent only upon polar effects, since it was argued that steric effects should approximately cancel in the difference of relative rates (3.28) (Ingold, 1969). In general, the precision of the fit of data to (3.27) is not as good as that for correlations involving the Hammett equation. Derivations are frequently observed when the point of substitution is close to the reaction centre, and when highly ramified

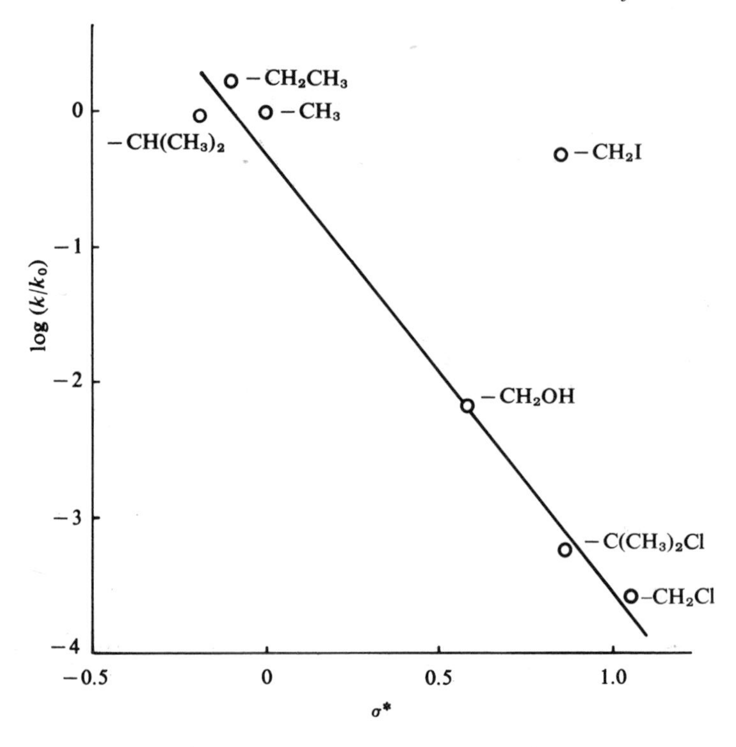

Fig. 3.6 ρ^* σ^* correlation for solvolysis of tertiary alkyl chlorides $R(CH_3)_2CCl$ in 80 per cent aqueous ethanol at 25 °C. The substituent attached to each point represents R. Data from Streitwieser (1956).

substituents are introduced, presumably since then steric factors begin to assume an important role.

The rates of solvolysis of many alkyl substrates have been correlated with σ^*-values (Streitwieser, 1956), and a selection of data for some tertiary systems is shown in fig. 3.6. The value of ρ^*, -3.29, implies the development of positive charge in the transition state, as expected for the S_N1 reactions of these tertiary systems; primary systems have ρ^*-values much closer to zero. The position of the point for R = $—CH_2I$ is seen to be markedly displaced from the correlation line. The β-I atom is in a position where it can interact with the reaction centre; the large rate enhancement (i.e. the rate in excess of that expected on the basis of a polar effect alone) can be attributed to anchimeric assistance (§5.2.2).

3.6.3. The mY correlation. Grunwald and Winstein (1948) defined the polarity (ionising power) of a solvent in terms of a parameter, Y (3.29) obtained from the rate constant of solvolysis, k, of t-butyl chloride in the solvent relative to its rate constant of solvolysis, k_0, in 80 per cent ethanol, which thus became the reference solvent ($Y = 0$). From this

$$\log (k/k_0)_{\text{t-BuCl}} = Y \tag{3.29}$$

definition, solvents in which the rate of solvolysis of t-butyl chloride is faster than that in the reference solvent have positive Y-values and such solvents are considered to be more polar than the reference solvent.

The rates of solvolysis of many substrates are found to be correlated with the Y-value of the solvent (3.30); m is the slope of the plot of log k

$$\log k = mY + \log k_0 \tag{3.30}$$

against Y and log k_0 is the intercept. Equation (3.30) has the form of a linear free energy relationship in which Y is the solvent parameter, and m measures the sensitivity of the substrate to the polarity of the solvent. However, it is subject to the important limitation that different values of m and log k_0 are required for each binary solvent system (Wells, 1968). This failure of (3.30) might be due either to the anomalous behaviour of the reference compound or to the inability of a simple linear free energy relationship to take into account different contributions of specific and general interactions between the solvent and the reacting substrate.

Anomalous behaviour in the reactions of t-butyl chloride might arise if the mechanism were not limiting S_N1, but required some nucleophilic participation by the solvent, or if elimination occurred concurrently with the solvolysis. These possibilities would appear to be unlikely, since the rates of solvolysis of 1-adamantyl bromide (71) in a wide range

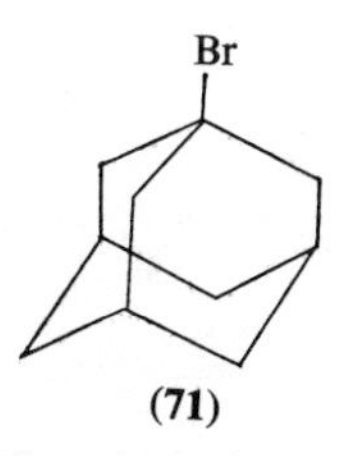

(71)

of solvents correlate well with Y-values (Raber, Bingham, Harris, Fry and Schleyer, 1970). Dispersion into separate correlations for binary solvent systems is indicated, but a reasonably good fit of most of the data to (3.30) is observed with $m = 1.20$ and log $k_0 = -6.45$. This shows that 1-adamantyl bromide (which undergoes solvolysis without solvent participation or elimination) and t-butyl chloride react by very similar mechanisms.

In the S_N1 mechanism of solvolysis the initial product of the ionisation might be best represented as an ion pair (§4.1.2) [*3.11*]. For such a

$$CH_3-\underset{CH_3}{\overset{CH_3}{C}}-Cl \underset{k_{-1}}{\overset{k_1}{\rightleftharpoons}} CH_3-\underset{CH_3}{\overset{CH_3}{C}}{}^{+}\ Cl^- \rightleftharpoons \begin{array}{c}\text{further dissociation}\\ \downarrow \\ \text{solvolysis product}\end{array} \qquad [3.11]$$

scheme, the rate of solvolysis will only be a true measure of the rate of ionisation (k_1) if no appreciable amount of the ion pair returns to the starting material, i.e. if k_1 and $k_{-1} \ll k$s of all the other steps. If ion-pair return is not negligible, then this might be responsible for part of the dispersion observed, if solvolysis in the different binary solvent systems is characterised by different amounts of ion-pair return.

Despite its limitations, the mY correlation is a useful empirical relationship that may be used to obtain fairly reliable estimates of solvolysis rate constants. In addition, the parameter m gives some information about the mechanism of solvolysis. A more polar transition state is involved in an S_N1 reaction than in an S_N2 reaction, so that we may expect that a compound which reacts by an S_N1 mechanism will be more sensitive to the ionising power of the solvent, and will be characterised by a larger value of m than one which reacts by an S_N2 mechanism (Sykes, 1972). In general, values of m close to unity indicate an S_N1 mechanism and those of *ca* 0.3 indicate an S_N2 mechanism; intermediate values ($\approx$0.5) may be taken as an indication of borderline behaviour.

4 Ion-pair intermediates

4.1. Introduction

The S_N1 mechanism of nucleophilic substitution was originally represented as the rate-limiting heterolysis of the substrate molecule to give a carbonium ion which then reacted rapidly with any available nucleophile. According to the present view, the heterolysis may give first an ion pair (R^+X^-) which on further dissociation leads to the fully dissociated, or 'free', carbonium ion [*4.1*]. The important feature of this

$$RX \underset{}{\overset{\text{ionisation}}{\rightleftharpoons}} (R^+X^-) \underset{}{\overset{\text{dissociation}}{\rightleftharpoons}} R^+ + X^- \qquad [4.1]$$

hypothesis is that ionisation and dissociation are regarded as being separate stages in the formation of a free carbonium ion; either the ionisation step, or one of the subsequent steps in the dissociation, may be rate limiting. Reaction with a nucleophile may occur at an ion-pair stage or at the free carbonium-ion stage, so that more than one intermediate is possible. Since more steps have to be considered in the ionisation and dissociation of a substrate than in the original scheme, the kinetic form of an S_N1 process can be more complex than that discussed in chapter 3 (§3.1). There is now a great deal of experimental evidence which requires that intermediates other than free carbonium ions be postulated in many S_N1 reactions (Winstein, Appel, Baker and Diaz, 1965). We shall discuss some of this evidence in the present chapter, but to begin with we shall consider ionic association and the nature of ion pairs.

4.1.1. Ionic association. The idea of ionic association has been used for many years to provide a relatively simple explanation of the, often complex, behaviour of electrolytes in solution. The conductance of an electrolyte may be expressed in terms of the Onsager limiting equation (4.1), in which Λ_0 is the conductance at infinite dilution, and α and β are constants which depend only on the dielectric constant and on the viscosity of the solvent. For dilute solutions ($c < 10^{-3}$ mol l^{-1}) of many univalent electrolytes in water, and in other solvents of high dielectric constant, the variation of the conductance, Λ, with the concentration of the electrolyte, c, is as predicted by (4.1). In solvents of low dielectric

$$\Lambda = \Lambda_0 - (\alpha\Lambda_0 + \beta)\sqrt{c} \tag{4.1}$$

constant, large deviations are often observed in which Λ decreases more rapidly than predicted. Such deviations are attributed to ionic association, which reduces the concentration of the current-carrying charged particles. Electrostatic attractions between ions of opposite charge lead to the formation of ionic aggregates in solution, these may be uncharged, e.g. (R^+X^-) and ($R^+X^-R^+X^-$), or carry only a single charge, e.g. ($X^-R^+X^-$) and ($R^+X^-R^+$). The importance of the different types of association will depend on the concentration and the nature of the electrolyte, and upon the nature of the solvent. For dilute solutions in solvents of dielectric constant greater than about 12, the formation of aggregates larger than ion pairs can usually be neglected.

4.1.2. The nature of ion pairs. According to the electrostatic theory of Bjerrum, the potential energy, E_r, of a pair of ions may be expressed as a function of the inter-ionic separation, r (4.2). In this equation, Z_R and

$$E_r = Z_R Z_X e^2/Dr \tag{4.2}$$

Z_X are the ionic charges, e is the electronic charge, and D is the dielectric constant. Using this potential energy expression it is possible to calculate the probability of finding an ion R at a distance r from an ion of opposite charge X. This treatment results in a distribution such that the probability is high for small values of r and for large values of r, but low for intermediate values of r. The probability has a minimum value at a separation, r_{min}, given by (4.3), in which k is Boltzmann's constant

$$r_{min} = |Z_R Z_X| e^2/2DkT \tag{4.3}$$

(Robinson and Stokes, 1959). At the separation of r_{min} the potential energy of the two ions is $2kT$, which represents the energy of their mutual electrostatic attraction. For separations greater than r_{min}, the thermal energy of the ions will be greater than the electrostatic energy, and the ions can be regarded as being free, i.e. completely dissociated. However, for separations smaller than r_{min} the electrostatic energy is greater than the thermal energy, and the ions are best regarded as a single species, i.e. as an ion pair. According to the Bjerrum theory, therefore, any pair of ions of opposite charge will constitute an ion pair when their inter-ionic separation is less than a certain critical value, r_{min}, which depends upon the dielectric constant of the solvent; some typical values for uni-univalent electrolytes are water (3.6 Å), acetic acid (45 Å) and benzene (120 Å).

The above treatment neglects the effect of solvation, and gives no indication of the structure of the ion pairs. In solvents of low dielectric constant the values of r_{min} are large enough to include many possibilities for the structure, from a pair of ions in contact, to a pair of ions separated by many solvent molecules. Whilst this level of definition is sufficient for many purposes it is often desirable, especially when considering reactions of carbonium ions, to be more explicit about the structure, or range of structures, of the ion pairs. In order to do this, let us consider the association of an anion, X^-, and a carbonium ion, R^+, in a medium for which the value of r_{min} is much greater than the sum of the ionic radii (Bethell and Gold, 1967).

For large separations, the ions will be independent of one another and both will be solvated. As they come together the potential energy of the system will be reduced by an amount equal to the electrostatic interaction (4.2) (fig. 4.1), until at some separation ($<r_{min}$) the decrease in E_r is greater than that expected from (4.2) and a shallow minimum is reached. At this stage, labelled $R^+ \parallel X^-$ in fig. 4.1, the ions are separated by only one or two solvent molecules, and the electrostatic attraction is large because of the dielectric saturation of the space between the ions. Further approach of the ions requires the removal of the solvent molecules from the space between the ions, a process which requires an appreciable amount of energy. Evidence for this last point comes from measurements of ultrasonic absorption by solutions of inorganic salts (Atkinson and Kor, 1965). The final stage is the pair of ions in contact, R^+X^-, within a common solvation shell. Further collapse, to the covalent molecule, RX, will require a small activation energy because the redistribution of charge will probably lead to some change in the structure of the solvation shell.

The above description is not intended to be rigorous, but merely to indicate the basis for the belief that various discrete structures may be assigned to ion pairs. Only two such structures have been considered explicitly, R^+X^-, a 'contact', or 'intimate' ion pair, and $R^+ \parallel X^-$, a 'solvent separated' ion pair. Of course there may be more than one energy barrier between these two species so that other discrete structures may exist (Szwarc, 1969).

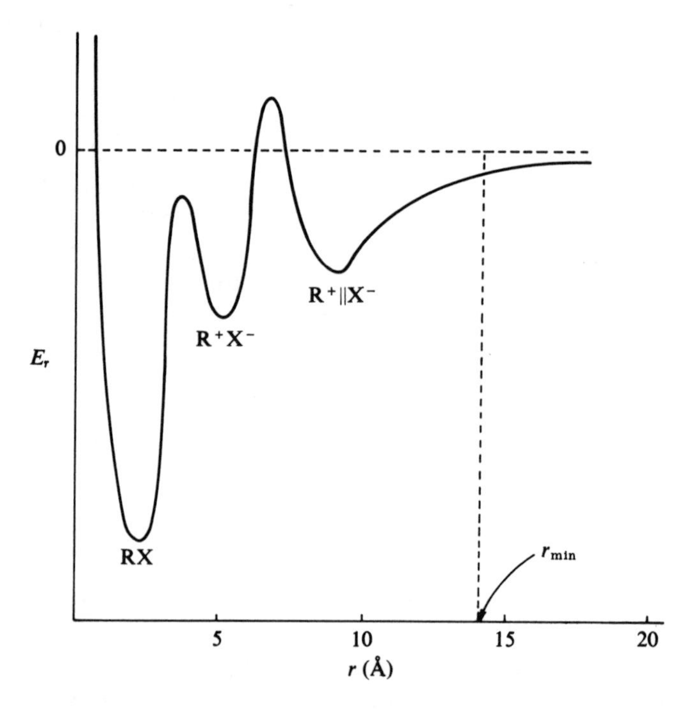

Fig. 4.1 A representation of the energy profile for the association of ions R^+ and X^- in a solvent of low dielectric constant.

By the principle of microscopic reversibility, the ionisation and dissociation of RX will follow the same course in reverse, so that the formation of a free carbonium ion from RX may be represented by [*4.2*]. The

$$RX \rightleftharpoons R^+X^- \rightleftharpoons R^+ \| X^- \rightleftharpoons R^+ + X^- \qquad [4.2]$$

preceding analysis only applies to solvents of low dielectric constant for which r_{min} is fairly large (> 10 Å). As the dielectric constant of the solvent increases so r_{min} becomes smaller, and a point may be reached when it is no longer meaningful to attribute discrete structures to the ion pairs, or indeed even to consider ion pairs at all.

4.2. Ion-pair intermediates in solvolytic nucleophilic substitutions

The solvolysis of optically active *exo*-2-norbornyl brosylate (72) in acetic acid and in aqueous acetone gives the corresponding *exo*-2-norbornyl product (73) with complete loss of optical activity (Winstein and Trifan, 1952). The first-order polarimetric rate constant, k_α, based

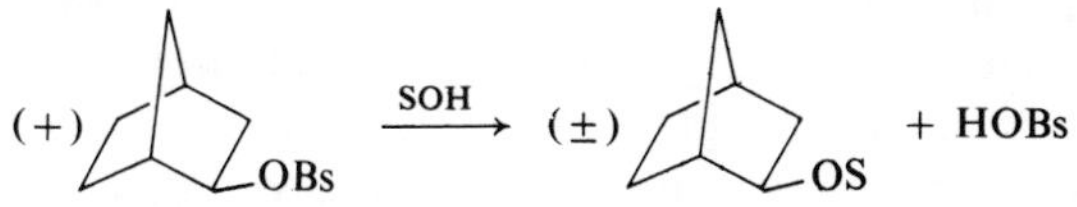

(+) [structure] OBs —SOH→ (±) [structure] OS + HOBs

(72) (73)

on the rate of loss of optical activity [*4.3*], was found to be larger than the first-order titrimetric rate constant, k_t, based on the rate of appearance of *p*-bromobenzenesulphonic acid [*4.4*]. This result may be understood

$$(+)\ \text{RX} \xrightarrow{k_\alpha} (\pm)\ \text{RX} + (\pm)\ \text{ROS} \qquad [4.3]$$

$$k_\alpha = (2.303/t) \log [\alpha_0/\alpha_t]$$

$$\text{RX} \xrightarrow{k_t} \text{ROS} + \text{HX} \qquad [4.4]$$

$$k_t = (2.303/t) \log [H_\infty/(H_\infty - H_t)]$$

stood if it is assumed that in both solvents part of the loss of optical activity that occurs during the solvolysis is due to the racemisation of the starting material [*4.3*]. The carbonium ion involved in the solvolysis of (72) is believed to have the symmetrical bridged structure (74) (cf. chapter 5), so that the recombination of the carbonium ion with the counter ion would necessarily lead to racemic starting material.

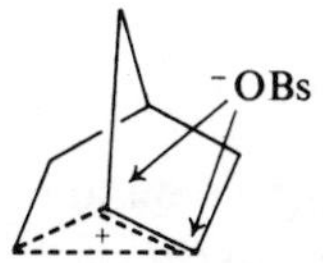

$^-$OBs

(74)

The important observation is that both the loss of optical activity and the production of acid are strict first-order kinetic processes; the racemisation of the starting material cannot therefore involve the free carbonium ion (74), and so the simple scheme [*4.5*] may be excluded. If

$$\text{RX} \rightleftharpoons \text{R}^+ + \text{X}^- \longrightarrow \text{ROS} \qquad [4.5]$$

racemisation of RX were the result of return to starting material from R^+, then k_α might be constant, but k_t would decrease with time because of the common-ion effect (§3.2.1). Likewise, if racemisation were due to attack of X^- on unreacted RX, then k_t might be constant, but k_α would increase with time.

The excess rate of loss of optical activity over rate of solvolysis can be explained in terms of return to covalent starting material from ionised but undissociated ion pairs [*4.6*], i.e. from an intermediate in which R^+ and X^- have not attained kinetic independence. Those ion pairs not

$$\mathrm{RX} \rightleftharpoons (\mathrm{R^+X^-}) \rightleftharpoons \mathrm{R^+} + \mathrm{X^-} \longrightarrow \mathrm{ROS} \qquad [4.6]$$

collapsing to covalent RX are assumed to separate further, possibly to free ions, and to give rise to solvolysis product.

4.2.1. The dual ion-pair hypothesis. Stereochemical results, of the type mentioned above, have played an important part in providing evidence for the involvement of ion-pair intermediates. In the acetolysis of *threo*-3-*p*-anisyl-2-butyl brosylate (75) the ratio of first-order rate constants, k_α/k_t, was found to be 4:1, which indicates that return from an ion-pair intermediate is important (Winstein and Robinson, 1958). There is much evidence to suggest that (75) ionises to a bridged carbonium ion (cf. chapter 5) so that the ion pair may be represented by (76). It will be

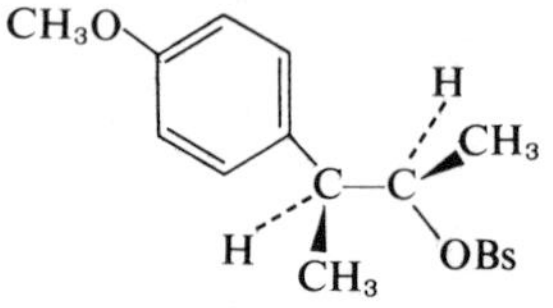

(75) 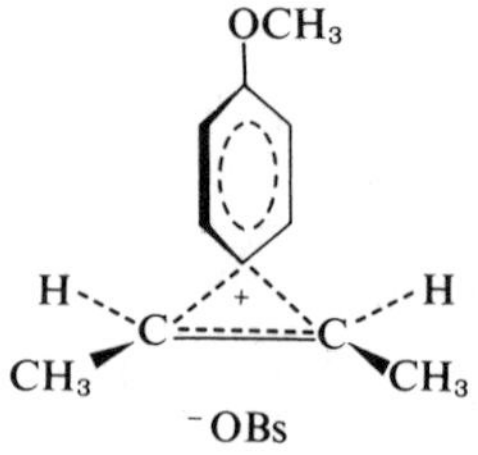 **(76)**

noted that this is a particularly favourable system with which to observe racemisation, since only a relatively small movement of the counter ion is necessary to produce a symmetrical intermediate. All products derived from (76) will be optically inactive; thus the polarimetric rate constant, k_α, provides a measure of the rate of ionisation of (75). Since ion-pair return results in racemisation of the starting material, the expression $(k_\alpha - k_t)$ gives the rate constant of racemisation [*4.7*].

$$(+)\,\mathrm{RX} \xrightarrow{k_{\mathrm{rac}}} (\pm)\,\mathrm{RX} \qquad [4.7]$$
$$k_{\mathrm{rac}} = k_\alpha - k_t$$

Additional information about the nature of the intermediates involved in the acetolysis of (75) has been obtained from a study of salt effects. With added lithium perchlorate (<0.1 mol l^{-1}) the value of k_α increases linearly with the concentration of salt (fig. 4.2) and follows the

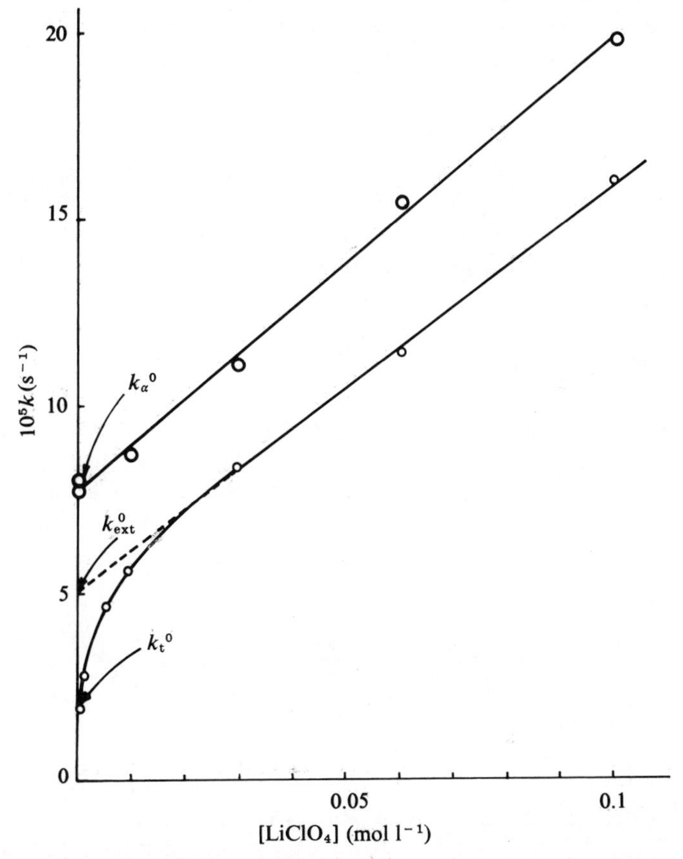

Fig. 4.2 The effect of added lithium perchlorate on the acetolysis of *threo*-3-*p*-anisyl-2-butyl brosylate. Data from Winstein and Robinson (1958).

normal pattern of behaviour for salt effects in acetic acid (§3.2) (4.4). In

$$k_\alpha = k_\alpha{}^0(1 + b_\alpha[\text{LiClO}_4]) \qquad (4.4)$$

this equation $k_\alpha{}^0$ is the polarimetric rate constant in the absence of added salt, and b_α is the slope of the line obtained by plotting k_α against $[\text{LiClO}_4]$.

A different type of behaviour is observed when k_t is plotted against $[LiClO_4]$. At low concentrations of salt (<0.02 mol l^{-1}) there is an initial rapid increase of k_t with increasing concentration of salt, the 'special salt effect', before the normal salt effect develops (fig. 4.2). The linear portion of the graph obeys (4.5), in which $k_{ext}{}^0$ is the value that the titrimetric rate constant would have at zero salt concentration in the absence of the special salt effect. The observed titrimetric rate constant

$$k_t = k^0_{ext}(1 + b_t[LiClO_4]) \qquad (4.5)$$

in the absence of added salt, $k_t{}^0$, is less than the extrapolated value of k^0_{ext}, and the ratio $k^0_{ext}/k_t{}^0$ (2.58 for the present system) provides a measure of the magnitude of the special salt effect. The fact that $k_\alpha > k_t$ for all concentrations of salt is taken to mean that ion-pair return is important under all conditions of acetolysis. The interesting feature of the results is that the special salt effect partly closes the gap between the polarimetric and the titrimetric rates, i.e. it apparently eliminates some of the ion-pair return.

These results have been rationalised in terms of the reaction scheme [*4.8*] which includes two kinds of ion pair, an intimate ion pair (77) and a solvent separated ion pair (78). The scheme shown includes the special

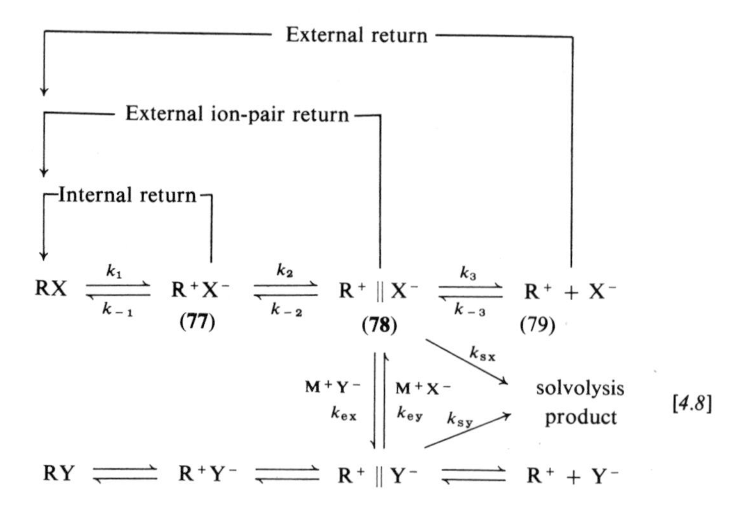

salt effect and gives the necessary terminology associated with the dual ion-pair hypothesis (Winstein, Klinedinst and Robinson, 1961).

In the general case, ionisation of RX gives the intimate ion pair (77) which on dissociation gives first the solvent separated ion pair (78) and finally the free carbonium ion (79). Nucleophilic attack by an external reagent may occur at any stage of the ionisation and dissociation of RX, but for the acetolysis of (75) the results seem to be quite consistent with the assumption that product arises only by capture of the solvent separated ion pair (78).

The special salt effect is explained by the suppression of external ion-pair return. The added salt, which under the experimental conditions is present largely in the form of ion pairs M^+Y^-, is assumed to undergo

$$R^+ \| X^- + M^+Y^- \rightleftharpoons R^+ \| Y^- + M^+X^- \qquad [4.9]$$

the metathesis reaction [*4.9*], which produces the new solvent separated ion pair $R^+ \| Y^-$. This exchange may lead, by ion-pair return, to new covalent substrate RY, but with such a weak nucleophile as the perchlorate ion such return is thought to be unlikely, and $R^+ \| Y^-$ most probably reacts rapidly to give solvolysis product. The special salt therefore effectively traps all of the solvent separated ion pair (78) and prevents external ion-pair return. Since an exchange similar to [*4.9*] does not apparently occur with the intimate ion pair (77), the special salt does not affect internal return, hence the ratio k_α/k_t is reduced by the special salt effect but not to unity.

By applying the steady state approximation to [*4.8*] for the case in which solvolysis product arises only from the solvent separated ion pair, we may obtain (4.6), in which k_{ext} is the extrapolated value of k_t (fig. 4.2); *a*, *b* and *c* are functions of the various rate constants shown in [*4.8*] (except for k_3 and k_{-3}). Equation (4.6) accounts quantitatively not

$$\frac{k_t}{k_{ext} - k_t} = a + \frac{b[M^+Y^-]}{1 + c[M^+X^-]} \qquad (4.6)$$

only for the special salt effect, but also for the effect of added common salt M^+X^-. For the reaction in question, the acetolysis of (75), the addition of lithium brosylate to the reaction solution in the presence of lithium perchlorate depresses the special salt effect by the amount predicted by (4.6). This result, called 'induced common ion rate depression', can be explained in terms of the effect of added salt M^+X^- on the exchange [*4.9*]. The addition of lithium brosylate alone produces a small normal salt effect; the common ion effect is not observed, which indicates that external return is not an important factor, and possibly that free carbonium ions are not involved in the acetolysis.

4.2.2. The ion-quadrupole hypothesis. The success of (4.6) in accounting for the observed salt effects in acetolysis reactions provides convincing support for the dual ion-pair scheme. However, the special salt effect can be explained by an alternative scheme involving only one ion-pair intermediate (80) and an ion quadrupole (81), [*4.10*] (Ingold, 1969).

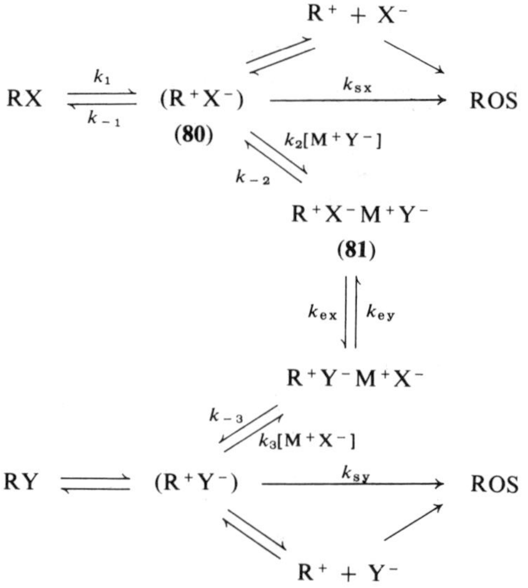

$$
\begin{array}{ccccc}
 & & R^+ + X^- & & \\
 & & \nearrow\!\!\swarrow & \searrow & \\
RX & \underset{k_{-1}}{\overset{k_1}{\rightleftharpoons}} & (R^+X^-)\ \mathbf{(80)} & \xrightarrow{k_{sX}} & ROS \\
 & & \underset{k_{-2}}{\overset{k_2[M^+Y^-]}{\rightleftharpoons}} & & \\
 & & R^+X^-M^+Y^-\ \mathbf{(81)} & & \\
 & & k_{eX} \Big\downarrow\!\!\Big\uparrow k_{eY} & & \\
 & & R^+Y^-M^+X^- & & \\
 & & \underset{k_3[M^+X^-]}{\overset{k_{-3}}{\rightleftharpoons}} & & \\
RY & \rightleftharpoons & (R^+Y^-) & \xrightarrow{k_{sY}} & ROS \\
 & & \searrow\!\!\nwarrow & \nearrow & \\
 & & R^+ + Y^- & &
\end{array}
\qquad [4.10]
$$

Application of the steady state approximation to scheme [*4.10*] leads to a kinetic expression of the same form as that of (4.6), for low concentrations of the special salt MY, although the constants *a*, *b* and *c* are now functions of the various rate constants shown in [*4.10*]. Therefore, both of the reaction schemes [*4.8*] and [*4.10*] lead to the same predictions concerning the behaviour of k_t in the region of the special salt effect.

There is, however, one important difference between the two schemes which, in principle, might allow them to be distinguished one from the other. As we have seen, according to scheme [*4.8*] the special salt effect is attributed to the suppression of external ion-pair return. When sufficient special salt has been added, k_t is expected to show a normal salt effect, since at this stage return from $R^+ \parallel X^-$ has been totally suppressed. This is a result of the effect of large concentrations of M^+Y^- on the exchange [*4.9*]; $R^+ \parallel Y^-$ is assumed to react rapidly and is therefore removed as fast as it is formed. Extrapolation of the linear part of the k_t curve (fig. 4.2) to zero salt concentration thus gives the value that k_t^0

would have in the absence of return from $R^+ \parallel X^-$, and it follows that all special salts should lead to the same value of k^0_{ext}.

According to [*4.10*] the ion pair exchange is assumed to occur by the formation of an ion quadrupole (81) which then undergoes rearrangement. The concentration of special salt can effect the relative proportions of (80) and (81), but it cannot affect significantly the ion-quadrupole rearrangement, and it need not suppress completely the step with rate constant k_{-2}. The linear part of the k_t curve thus measures a normal salt effect on which might be superimposed a residual and constant amount of ion-pair return from (81). Since this residual amount of return need not be the same for the different special salts, extrapolation to zero salt concentration need not yield the same value of k^0_{ext}. The distinction to be made between the alternative schemes [*4.8*] and [*4.10*] therefore rests upon whether the different special salts give different values for k^0_{ext}, and upon whether these differences are large enough to be determined experimentally. Although data are available for several reactions it is not possible, at present, to distinguish between the alternative schemes.

4.2.3. Factors influencing dissociation. As already mentioned, nucleophilic attack by an external reagent may occur at any stage during the ionisation and dissociation of a given substrate. In terms of scheme [*4.8*] the intermediates necessarily involved in any given reaction will depend on both the nature of the substrate and the nature of the solvent. It might be expected that as the structure of the substrate increasingly favours the formation of a stable carbonium ion, and as the ability of the solvent to support ionic dissociation increases, so the probability increases that fully dissociated ions will be involved as reaction intermediates. Information concerning the degree of dissociation reached in many acetolysis reactions has been obtained from studies of salt effects.

In table 4.1 are summarised some of the results that have been obtained for alkyl arenesulphonates of the 3-phenyl-2-butyl type of structure; the compounds are arranged in decreasing order of stability of the corresponding carbonium ions. The observation of a common-ion effect with 2-(2,4-dimethoxyphenyl)ethyl brosylate indicates that free carbonium ions are involved in the acetolysis of this substrate, since common-ion depression is diagnostic of external return. However, it is found that there is a limit to the extent to which common-ion salt can depress the rate of acetolysis, a minimum rate is reached at relatively low concentrations of added lithium brosylate, and thereafter a small normal salt effect is observed (Winstein, Clippinger, Fainberg, Heck and

Robinson, 1956). Such an effect is in marked contrast to the normal common-ion effect (§3.2.1), and it is only possible if the common ion returns the dissociated carbonium ion to a species which can itself give rise to solvolysis product; in the present case this species appears to be the solvent separated ion pair. After sufficient common-ion salt has

TABLE 4.1 *Salt effects on the acetolysis of some alkyl arenesulphonates at 50 °C*[a]

Compound	b_t[b]	Common ion depression	k^0_{ext}/k_t^0
2-(2,4-dimethoxyphenyl)ethyl OBs	12	Yes	2.2
threo-3-*p*-anisyl-2-butyl OBs[c]	18	No	2.6
1-*p*-anisyl-2-propyl OTs	27	No	2.4
threo-3-phenyl-2-butyl OTs	37	No	1.0

[a] Data from Winstein and Clippinger (1956).
[b] Normal salt effect parameter cf. (4.5) for $LiClO_4$. [c] 25 °C.

been added to cause complete return of the free carbonium ion, it can only influence the rate of reaction by a normal salt effect on the ionisation step.

A special salt effect is observed in the acetolysis of 2-(2,4-dimethoxyphenyl)ethyl brosylate ($k^0_{ext}/k_t^0 = 2.2$), which indicates that external ion-pair return is an important factor. The salt effects observed with this substrate therefore indicate that ionisation and dissociation produce the free carbonium ion, that capture by the solvent occurs at the solvent separated ion-pair and free carbonium-ion stages, and that return from both free carbonium ions and intermediate ion pairs is important.

The acetolysis of *threo*-3-*p*-anisyl-2-butyl brosylate does not show a common-ion effect, a result usually taken to mean that free carbonium ions are not involved (however, cf. Winstein, Clippinger, Fainberg, Heck and Robinson, 1956). A special salt effect is observed, however, and this indicates that dissociation reaches at least the solvent separated ion-pair stage. As mentioned in an earlier section, the results are consistent with product arising only by capture of the solvent separated ion pair, and ion-pair return involves both the intimate and the solvent separated ion pairs.

In the case of *threo*-3-phenyl-2-butyl tosylate the absence of the common-ion effect again indicates that free carbonium ions are not involved in the acetolysis. The absence of a special salt effect also indicates that return from the solvent separated ion pair is not important. Ion-pair return (from the intimate ion pair) does occur, however, as indicated by the fact that the polarimetric rate exceeds the titrimetric rate, $k_\alpha{}^0/k_t{}^0 = 4.6$. In this case solvolysis product is thought to arise by solvent capture of the solvent separated ion pair, but return involves only the intimate ion pair. The preceding examples clearly illustrate how the degree of dissociation, and the nature of return, are dependent on the structure of the reacting substrate. The solvent is also important, but its effect is more difficult to predict because the nucleophilicity, ionising power, and ability to dissociate ion pairs all have to be taken into account.

The solvolysis of *threo*-3-*p*-anisyl-2-butyl brosylate has been studied in a number of solvents, and the results (table 4.2) indicate that the amount of return decreases as the ionising power of the solvent increases. A measure of the latter quantity is provided by the value of $k_\alpha{}^0$,

TABLE 4.2 *The solvolysis of threo-3-p-anisyl-2-butyl brosylate at 25°C in various solvents*[a]

Solvent	$k_\alpha{}^\circ$ (rel)	$k_\alpha{}^\circ/k_t{}^\circ$	Total return (%)[b]
12.5% acetic acid–dioxane	1.0	20	95
10% acetic acid–benzene	1.6	16	94
10% formic acid–dioxane	7.2	16	94
Ethanol	80	1.27	21
Acetic acid	235	4.1	75
25% formic acid–acetic acid	1930	1.1	7

[a] Data from Winstein and Robinson (1958).
[b] Calculated from $100[1 - (k_t{}^\circ/k_\alpha{}^\circ)]$.

since for the present substrate k_α can be identified with the ionisation rate constant k_1 of scheme [*4.8*]. Although there is a good correlation between the ionising power of the solvent and the amount of return for most of the solvents listed, the result in ethanol shows that the nucleophilicity of the solvent cannot be ignored. A nucleophilic solvent is able to trap ion-pair intermediates efficiently and so reduce the amount of ion-pair return.

The importance of making the distinction between the ionising power of a solvent and its ability to dissociate ion pairs is made apparent by the results obtained for the acetolysis of *threo*-3-*p*-anisyl-2-butyl brosylate as acetic anhydride is added to the solvent (table 4.3). As the solvent is changed from pure acetic acid to pure acetic anhydride the ionising power (measured by $k_\alpha{}^0$) decreases smoothly. The titrimetric rate constant, $k_t{}^0$, passes through a maximum, however, which corresponds to a minimum in the ion-pair return ($k_\alpha{}^0/k_t{}^0$). The importance of the special salt effect ($k^0_{ext}/k_t{}^0$) gradually diminishes, while at the same time common-ion rate depression, which is not observed in acetic acid, gains in

TABLE 4.3 *The effect of added acetic anhydride on the acetolysis of threo-3-p-anisyl-2-butyl brosylate*[a]

	AcOH (25 °C)	50% AcOH–Ac_2O (25 °C)	Ac_2O (50 °C)
$10^5 k_\alpha{}^0$	7.98	5.78	4.42
$10^5 k_t{}^0$	1.96	3.18	1.07
$k_\alpha{}^0/k_t{}^0$	4.1	1.8	4.1
$k^0_{ext}/k_t{}^0$	2.58	1.56	1.47
% product from solvent separated ion pair	100	21	3
% product from dissociated ion	0	79	97

[a] Data from Winstein, Appel, Baker and Diaz (1965).

importance. These results can be explained by assuming that whereas acetic acid is the better ionising medium, acetic anhydride is the more dissociating. Thus in acetic acid solvolysis product arises from the solvent separated ion pair, and ion-pair return is important. In acetic anhydride product arises from the free carbonium ion, and return is not so important. Acetic anhydride reduces the rate of formation of ionic intermediates but increases the extent to which they are dissociated.

4.2.4. Racemisation, equilibration and ion-pair return. The racemisation of the starting material that often accompanies solvolysis has been attributed to ion-pair return (§4.2.1), and the rate constant for racemisation, k_{rac}, therefore provides a measure of the importance of that return. For arenesulphonate and benzoate esters an independent estimate of ion-pair return is available, since ionisation of the starting material results in the randomisation of the oxygen atoms, and ion-pair return leads to equilibration of the oxygen atoms in the starting material [*4.11*]. In this equation k_{eq} is the first-order rate constant for ^{18}O-equilibration, and a refers to the ^{18}O-content of the alcohol obtained after

$$ROS^{18}O_2Ar \xrightarrow{k_{eq}} R^{18}OS^{18}O_2Ar \qquad [4.11]$$

$$k_{eq} = (2.303/t) \log [(a_\infty - a_0)/(a_\infty - a_t)]$$

cleaving the S–O linkage of the ester; the subscripts have their usual meaning. It should be emphasised that when using racemisation, or equilibration, to measure ion-pair return we make the implicit assumption that these processes are an integral part of the solvolysis reaction; this assumption might not always be valid (Hammett, 1970). Furthermore, in the general case, racemisation and equilibration provide only lower estimates of ion-pair return, because return with either retention of configuration or incomplete randomisation cannot be detected.

Oxygen equilibration and racemisation occur in the unreacted ester during the acetolysis of both *threo*-3-phenyl-2-butyl tosylate (82) and *endo*-bicyclo[3.2.1]octan-2-yl tosylate (83). In both cases $k_{eq}/k_{rac} \approx 0.5$,

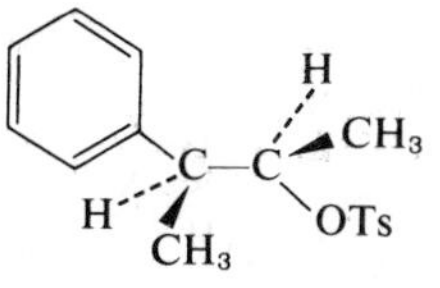

H CH3 OTs H CH3
(82)

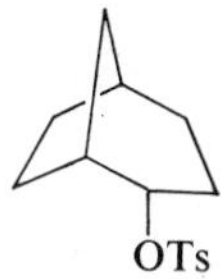

OTs
(83)

i.e. a substantial amount of return occurs without equilibration of the oxygen atoms (Goering and Thies, 1968). This result is almost certainly a consequence of the involvement of symmetrical bridged intermediates, cf. (74) and (76), in these solvolyses. Only a small change in the relative positions of the counter ions is required for return to lead to enantiomeric starting material, and such a change may occur whilst the ions are subject to strong interactions which prevent significant oxygen equilibration (84).

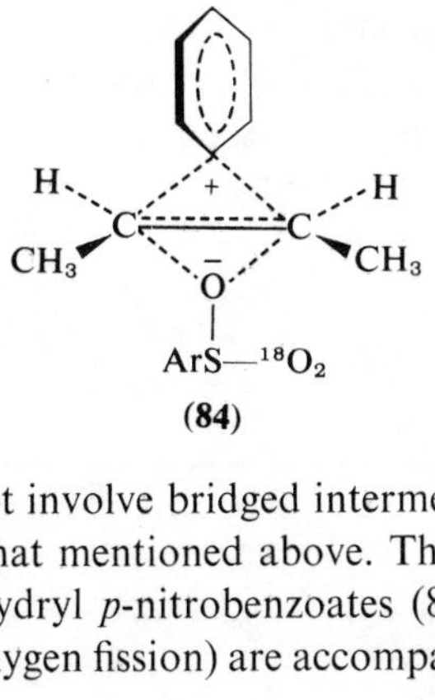

H + H CH3 C C CH3 O ArS—$^{18}O_2$
(84)

Systems which do not involve bridged intermediates show behaviour of a type different to that mentioned above. The hydrolyses of several *para*-substituted benzhydryl *p*-nitrobenzoates (85) in aqueous acetone (which involve alkyl–oxygen fission) are accompanied by ion-pair return

(Goering and Hopf, 1971). As the substituent, —Y, changes from —Cl to —OCH_3 (table 4.4) the rate of hydrolysis (k_t) increases by a factor of

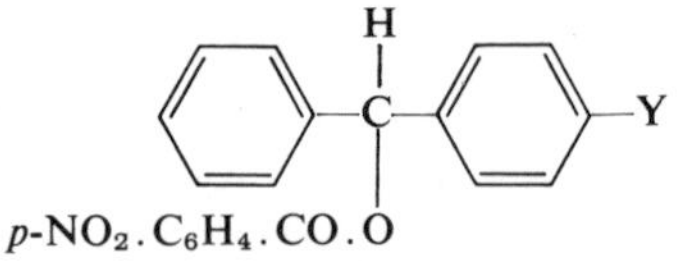

H

C —Y

$p\text{-}NO_2.C_6H_4.CO.O$

(85)

>2600; the amount of return, however, remains fairly constant, although the amount of racemisation associated with return (k_{rac}/k_{eq}) gradually increases. In all cases $k_{rac}/k_{eq} < 1$, i.e. a substantial amount of return occurs with retention of configuration. For such systems, k_{eq} affords a better measure of total ion-pair return than does k_{rac}.

TABLE 4.4 *The hydrolysis of* para-*substituted benzhydryl* p-*nitrobenzoates in 90 per cent aqueous acetone at 99.5 °C*[a]

Substituent	$10^3k_t(h^{-1})$	$10^3k_{eq}(h^{-1})$	Return (%)[b]
p-Cl	0.5	1.27	72
H	1.01	2.97	75
p-CH_3	11.20	32.3	74
p-OCH_3	>1300	>3300	72

[a] Data from Goering and Hopf (1971).
[b] Calculated from $100\ k_{eq}/(k_{eq} + k_t)$.

A similar correlation between reactivity and stereochemistry of return has also been noticed in the hydrolyses of 1-phenylethyl *p*-nitrobenzoate (86) and 1-*p*-anisylethyl *p*-nitrobenzoate (87) in 70 per cent aqueous acetone. In the case of (86), return occurs without racemisation, but with (87), which is more than 30 000 times more reactive than (86), return is accompanied by substantial racemisation, $k_{rac}/k_{eq} = 0.71$.

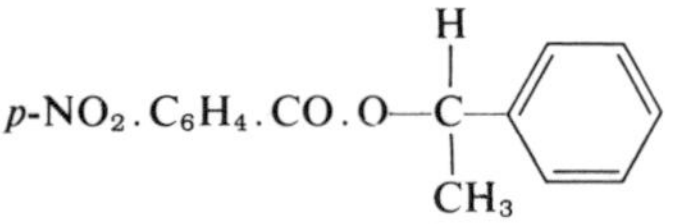

H | H
$p\text{-}NO_2.C_6H_4.CO.O—C—$ | $p\text{-}NO_2.C_6H_4.CO.O—C—$ —OCH_3
CH_3 | CH_3

(86) 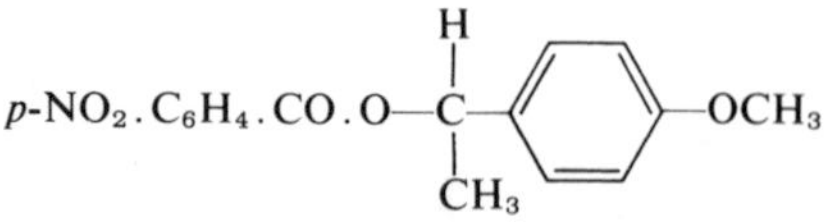 **(87)**

These results suggest that the interactions between the counter ions that are responsible for the retention of configuration become weaker as the carbonium ion becomes more stable, i.e. as the charge on the carbonium ion becomes more delocalised. An explanation in terms of the dual ion-pair scheme might be that more return occurs from the solvent separated ion pair with the more stable carbonium ions.

The effect of added sodium azide on the stereochemistry of ion-pair return also supports this idea. In the hydrolysis of *p*-chlorobenzhydryl *p*-nitrobenzoate in 90 per cent aqueous acetone, the addition of sodium azide (0.14 mol l^{-1}) eliminates racemisation but it still allows some ^{18}O-equilibration to occur. In the hydrolysis of *p*-methylbenzhydryl *p*-nitrobenzoate, on the other hand, added sodium azide decreases the amount of return, but it has little effect on the stereochemistry.

4.2.5. Secondary deuterium isotope effects. In an earlier section (§3.4) it was shown that the magnitude of the secondary deuterium isotope effect can be used to distinguish between solvolysis reactions that occur by the S_N1 and S_N2 mechanisms. We have seen in the present chapter, however, that an S_N1 process may in fact involve ion-pair intermediates, and the rate-limiting step of such a reaction may be the initial ionisation, or one of the subsequent steps in the further dissociation of the counter ions. It is therefore of interest to know what information the magnitude of the secondary deuterium isotope effect can give regarding the nature of the rate-limiting step.

For a reaction scheme such as [*4.12*], in which solvolysis product arises only from the solvent separated ion pair and $k_3 \gg k_{-2}$, it can be shown by using the steady state approximation that $k_{obs} = k_1F$, where

$$\mathrm{RX} \underset{k_{-1}}{\overset{k_1}{\rightleftharpoons}} \mathrm{R^+X^-} \underset{k_{-2}}{\overset{k_2}{\rightleftharpoons}} \mathrm{R^+ \| X^-} \xrightarrow{k_3} \text{product} \qquad [4.12]$$

$F = k_2/(k_{-1} + k_2)$. The term therefore describes the partitioning of the intimate ion pair between dissociation and return. A secondary deuterium isotope effect, calculated in terms of the observed solvolysis rate constants, may thus be represented by (4.7). The observed isotope effect

$$(k_H/k_D)_{obs} = (k_H/k_D)_1(F_H/F_D) \qquad (4.7)$$

is the product of the effects of isotopic substitution on the rate of ionisation and on the partitioning of the intimate ion pair.

An estimate of the partition isotope effect involved in the ethanolysis of benzhydryl benzoate has been made by studying the partitioning of benzhydryl benzoate ion pairs generated from diphenyldiazomethane and benzoic acid in ethanol (Murr and Donnelly, 1970) [*4.13*]. The titrimetric isotope effect $(k_H/k_D)_{obs}$ was found to be 1.19 at 25 °C, and from

$$\begin{array}{ccc} (C_6H_5)_2CN_2 + HOBz & \longrightarrow & (C_6H_5)_2CH\overset{+}{N}_2\overset{-}{O}Bz \\ & & \big\downarrow -N_2 \\ (C_6H_5)_2CHOBz & \rightleftharpoons & (C_6H_5)_2\overset{+}{C}H\overset{-}{O}Bz \\ & & \updownarrows \\ BzO^- = C_6H_5.CO_2^- & & (C_6H_5)_2\overset{+}{C}H \,\|\, \overset{-}{O}Bz \\ & & \big\downarrow \\ & & \text{product} \end{array} \quad [4.13]$$

the partitioning studies (F_H/F_D) was found to be 1.06. Similar values were estimated for the partition isotope effects of several other substrates. From these figures the isotope effect on the ionisation $(k_H/k_D)_1$ is calculated by (4.7) to have the value 1.12. In terms of this analysis a minimum solvolytic isotope effect is to be expected when the ionisation (k_1) is rate-limiting, and for such a reaction, product will arise either by solvent capture of the intimate ion pair, or by scheme [*4.12*] in which $k_2 \gg k_{-1}$ and $k_3 \gg k_{-2}$. A maximum isotope effect will be observed when the dissociation of the intimate ion pair (k_2) is rate-limiting and $k_{-1} \gg k_2$; in this case the partition isotope effect contributes to the observed solvolytic isotope effect (4.7). In the general case the secondary deuterium isotope effect may have any value between the limits set by (4.7), the actual value depending upon the relative magnitudes of the various rate constants.

The solvolyses of isopropyl arenesulphonates have been studied in several solvents. In trifluoroacetic acid the value of the secondary α-D effect, 1.22, is close to the maximum value (§3.4) and this has been interpreted in terms of a mechanism which involves the rate-limiting dissociation of an intimate ion pair (Shiner, 1970). In 70 per cent aqueous trifluoroethanol, in acetic acid, and in water, the value of the α-D effect is close to 1.14, while in 80 per cent aqueous ethanol it is 1.098. One explanation of the smaller-than-maximum α-D effect is to suppose that part of the solvolysis product arises by a direct S_N2 displacement on the covalent substrate. However, it would then be difficult to explain why the three solvents aqueous trifluoroethanol, acetic acid, and water, which differ markedly in their polarities and nucleophilicities, should give the same value for k_H/k_D. It seems preferable to assume that the value of 1.14 corresponds to the value for the α-D effect of an S_N1 process in which formation of the intimate ion pair is rate-limiting. This value agrees closely with that calculated (1.15) from the observed value, 1.22 in trifluoroacetic acid, assuming a value of 1.06 for the partition isotope effect. The value of 1.098 obtained for the α-D effect in 80 per cent aqueous ethanol is then probably due to an S_N2 component of the reaction.

4.3. The mechanisms of solvolytic nucleophilic substitutions

The important feature of the original classification of nucleophilic substitution was that two fundamentally distinct mechanisms were distinguished. The characteristics associated with these two mechanisms are sufficiently different, in extreme cases, for this distinction to provide a useful method of classification, although borderline reactions do create problems. Part of the difficulty arises because borderline reactions are frequently considered to be concurrent S_N1 and S_N2 processes, but the characteristics associated with the extremes of S_N1 and S_N2 behaviour probably do not apply in the borderline region. We have already mentioned (§1.3) that the characteristics of the S_N2 mechanism may show a considerable variation, depending upon the nature of the transition state involved in the substitution process. The discussion of the present chapter indicates that a variation is also possible in the characteristics associated with the S_N1 mechanism. It is likely that in the borderline region the differences in behaviour between the S_N1 and S_N2 mechanisms become so small that for practical purposes the mechanisms become indistinguishable. Whilst such a situation obviously detracts from the usefulness of the duality of mechanism as a means of classification, it in no way vitiates the fundamental distinction to be made between S_N1 and S_N2. No matter how the mechanism of a substitution reaction is described, each act of substitution must always involve either a bimolecular or a unimolecular pathway.

4.3.1. Necessary ion-pair intermediates. For a reaction which involves ionisation of the substrate, product may arise through interaction of the solvent with either the intimate ion pair, the solvent separated ion pair, or the fully dissociated carbonium ion [*4.14*]. In those cases which have

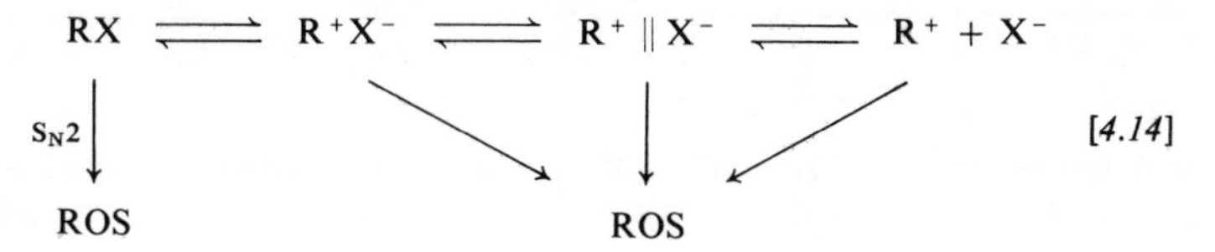

[*4.14*]

been studied in detail it has not been necessary to consider all of the various routes to product. The intermediates necessarily involved will depend upon the structure of the substrate and the nature of the solvent (§4.2.3). Even though the evidence may suggest that solvolysis product arises only from one type of intermediate, it may still be necessary to include additional intermediates in the reaction scheme in order to account for phenomena associated with ion-pair return (§4.2.4).

It is believed that the different ion pairs involved in a solvolysis scheme have discrete structures, although the actual structures are usually unknown. It is important, therefore, to remember that although the terms 'intimate' and 'solvent separated' are used to describe supposedly different ion-pair intermediates, these terms do not at present have precise structural significance. The need to emphasise this point arises because the dual ion-pair hypothesis, which was originally proposed to explain certain phenomena observed in acetic acid, is now applied to many different phenomena in other solvents. Although ion pairs may be required in these cases, they may well be different in structure from those intermediates of similar name involved in reactions in acetic acid. There is also the additional problem that in solvents of high dielectric constant it may not always be meaningful to consider ion pairs as reaction intermediates.

4.3.2. Tertiary systems. In this and the following sections some specific examples will be described in detail to illustrate the different types of behaviour which are thought to be due to ion-pair intermediates.

The behaviour of triphenylmethyl benzoate (88) in acetone indicates that even with substrates capable of forming carbonium ions as stable as

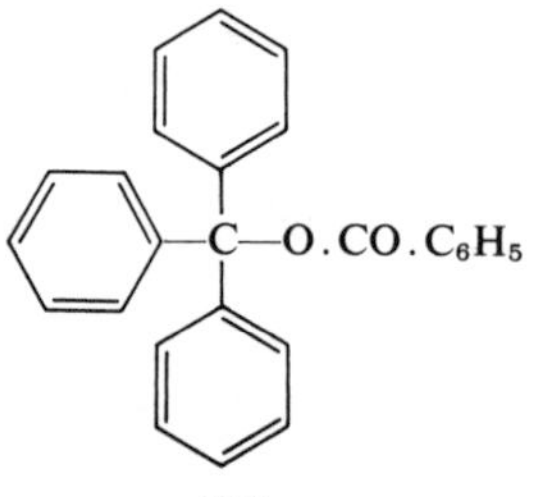

$(C_6H_5)_3C—O.CO.C_6H_5$

(88)

the triphenylmethyl cation, ion pairs cannot always be excluded (Winstein, Appel, Baker and Diaz, 1965). In anhydrous acetone the rate of ^{18}O-equilibration [*4.15*] is much greater than the rate of exchange with tetrabutylammonium [*carboxyl*-^{14}C] benzoate [*4.16*]. The rate

$$(C_6H_5)_3COC^{18}O.C_6H_5 \xrightarrow{k_{eq}} (C_6H_5)_3C^{18}OC^{18}O.C_6H_5 \qquad [4.15]$$

$$(C_6H_5)_3COCO.C_6H_5 + Bu_4\overset{+}{N}\ \overset{-}{O}{}^{14}CO.C_6H_5 \xrightarrow{k_e} (C_6H_5)_3CO^{14}CO.C_6H_5 + Bu_4\overset{+}{N}\ \overset{-}{O}CO.C_6H_5 \qquad [4.16]$$

constants for both reactions increase with an increase in the total concentration of salt (table 4.5). At the same total salt concentration the rate constant for consumption of azide ion is the same as that for benzoate exchange, and neither depends on the concentration of the nucleophile (Bu_4NOBz or Bu_4NN_3). Whilst the added azide suppresses the exchange reaction, it has little effect on the ^{18}O-equilibration.

The results thus indicate that the rate of ionisation exceeds the rate of chemical capture, and that ^{18}O-equilibration proceeds by way of an intermediate which is not easily captured by added salt. The exchange reaction and the capture by azide ion therefore occur with some other intermediate. The first intermediate appears to have the characteristics of an intimate ion pair, and the second intermediate could be either an additional ion pair or even the dissociated carbonium ion.

TABLE 4.5 *Salt effects on the reactions of triphenylmethyl benzoate in anhydrous acetone at 75 °C*[a]

10^2 [Salt] (mol l^{-1})		10^2 [Total salt] (mol l^{-1})	$10^6 k_{eq}$(s^{-1})[b]	$10^6 k$(s^{-1})[c]
–		–	4.6	
Bu_4NClO_4	(1.00)	1.00	7.2	
Bu_4NOBz	(0.039)	0.039		0.11
	(0.122)	0.122		0.17
	(0.30)	0.30		0.25
	(0.1–0.3)	6.00[d]		1.04
Bu_4NN_3	(0.3–1.1)	6.00[d]		0.97

[a] Data from Winstein, Appel, Baker and Diaz (1965).
[b] First-order rate constant for ^{18}O-equilibration.
[c] First-order rate constant for either benzoate exchange or formation of azide.
[d] Total salt concentration achieved by adding Bu_4NClO_4.

4.3.3. Secondary systems giving stable carbonium ions. The reactions of several secondary systems have already been described in detail in this chapter, and indeed the behaviour of such systems has provided much of the evidence upon which the ion-pair hypothesis is based. Thus compounds such as *threo*-3-*p*-anisyl-2-butyl brosylate have provided much information about solvolysis reactions in acetic acid, whilst compounds of the benzhydryl type have been useful in giving information about hydrolysis reactions in aqueous acetone.

The compound 1-phenylethyl chloride has long been considered a typical example of a secondary system which undergoes solvolysis by a limiting or S_N1 mechanism. The sensitivity of the rate of reaction to changes of the solvent, and to the nature of substituents in the benzene ring supports this conclusion. Additional evidence comes from measurements of secondary α-deuterium isotope effects, which show the maximum values expected for limiting behaviour (Shiner, Buddenbaum, Murr and Lamaty, 1968).

Stereochemical studies have revealed that, contrary to the generally held belief, the solvolysis of 1-phenylethyl chloride is accompanied by extensive inversion of configuration (Okamoto, Uchida, Saito and Shingu, 1966). Over the range of composition 50–80 per cent aqueous ethanol, the ethyl ether shows 34–42 per cent inversion; and the 1-phenylethyl alcohol shows 25–32 per cent inversion; even in water the alcohol is produced with 22 per cent inversion. One explanation of this result is to assume that the racemic product arises by solvent capture of the free carbonium ion, and that the inverted product arises by attack of the solvent on an ion pair. However, the almost constant amount of inversion observed as the composition of the solvent varies indicates that free carbonium ions are not involved, since we might then expect to observe more racemic products in the more aqueous solvents. The simplest assumption is that the product arises by capture of the solvent separated ion pair, and that free carbonium ions are not involved; the absence of a common-ion effect shows that return from the free carbonium ion is not important.

4.3.4. Simple secondary systems. Secondary systems that give rise to simple alkyl carbonium ions invariably exhibit borderline behaviour in their nucleophilic displacement reactions. Thus 2-octyl arenesulphonates (89) give substitution products with predominant, and in many cases total, inversion of configuration, although there is evidence that at least some of the products arise by a mechanism that involves ionisation. In the acetolysis of 2-octyl *p*-nitrobenzenesulphonate (RONs), for example, the addition of lithium tosylate leads to the formation of some

$$C_6H_{13}-\overset{\overset{\displaystyle CH_3}{|}}{\underset{\underset{\displaystyle H}{|}}{C}}-OSO_2.Ar$$

(89)

2-octyl tosylate by a process which seems best described as an ion-pair exchange reaction followed by ion-pair return [*4.17*] (Streitwieser and

$$\begin{array}{lllll}
RONs & \rightleftharpoons & R^+ONs^- & \longrightarrow & \text{solvolysis product} \\
 & & & \searrow\!\!\!\nwarrow & \\
 & & & & Li^+OTs^-\ R^+ONs^- \\
 & & & & \updownarrow \\
 & & & & OTs^-R^+ONs^-Li^+ \\
 & & & \nearrow\!\!\!\swarrow & \\
ROTs & \rightleftharpoons & R^+OTs^- & \longrightarrow & \text{solvolysis product}
\end{array} \quad [4.17]$$

Walsh, 1965). More direct evidence for ion-pair return is the ^{18}O-equilibration that occurs in the unreacted substrate during the solvolysis of 2-octyl brosylate (Diaz, Lazdins and Winstein, 1968*a*). The equilibration, $100\ k_{eq}/(k_{eq} + k_t)$, in the several solvents studied was methanol 1.1, acetic acid 6.5, formic acid 8.1, and trifluoroacetic acid 19.9.

These figures suggest that ion-pair return is not very important with simple secondary systems; however, it is possible that a large amount of return could have occurred before complete equilibration of the oxygen atoms. Evidence that this might be so was provided by a novel experiment which demonstrated that extensive return probably accompanies the solvolysis of isopropyl brosylate in trifluoroacetic acid (Shiner and Dowd, 1969). Isopropyl brosylate was found to react quantitatively in trifluoroacetic acid to give isopropyl trifluoroacetate, the half-life for the reaction being 182 minutes. Propene and *p*-bromobenzenesulphonic acid reacted together completely within 1 minute in trifluoroacetic acid to produce, in quantitative yield, isopropyl brosylate. (Propene was converted relatively slowly by trifluoroacetic acid alone, $t_{1/2} \approx 300$ minutes, into isopropyl trifluoroacetate.) It was also found that isopropyl alcohol and *p*-bromobenzenesulphonic acid reacted together quantitatively in trifluoroacetic acid to give isopropyl trifluoroacetate with a half-life of 7 minutes; no isopropyl brosylate was detected. The observations may be explained by reference to scheme [*4.18*].

$$\begin{array}{ccccccc} \mathrm{ROBs} & \underset{k_{-1}}{\overset{k_1}{\rightleftharpoons}} & \mathrm{R^+OBs^-} & \underset{k_{-2}}{\overset{k_2}{\rightleftharpoons}} & \mathrm{R^+ \| OBs^-} & \xrightarrow{k_3} & \text{product} \\ & & \uparrow & & \uparrow & & \\ & & \text{propene} + \mathrm{HOBs} & & \text{isopropyl alcohol} + \mathrm{HOBs} & & \end{array} \quad [4.18]$$

It was assumed that undissociated *p*-bromobenzenesulphonic acid (HOBs) reacted with propene to give initially an ion pair which would be a good representation of the intimate ion pair involved in the solvolysis reaction. This apparently collapses to covalent isopropyl brosylate faster than it dissociates, i.e. $k_{-1} \gg k_2$. Isopropyl alcohol and HOBs were assumed to lead to an ion pair of isopropyl cation and brosylate anion separated by a molecule of water, which was taken as a suitable model for the solvent separated ion pair. The intermediate formed by the addition, and therefore presumably $R^+ \| OBs^-$, reacts more rapidly to give solvolysis product than it returns to the intimate ion pair, i.e. $k_3 \gg k_{-2}$. These results therefore indicate that there is extensive return from the intimate ion pair, and that the rate-limiting step in the solvolysis is the dissociation of the intimate ion pair.

The addition of a strong nucleophile can also provide information about the role of intermediates in solvolysis reactions (§3.2.2). For example, the hydrolysis of 2-octyl methanesulphonate in 30 per cent aqueous dioxane containing sodium azide gives 2-octanol and 2-octyl azide, both products being formed with predominant inversion of configuration. The stereochemistry suggests an S_N2 mechanism, as does the fact that both the product composition and the rate of reaction are dependent upon the concentration of sodium azide. However, an analysis of the data has apparently indicated that quantitatively the kinetic results are not consistent with an S_N2 reaction (Sneen and Larsen, 1969). The results were explained in terms of scheme [*4.19*], in which the products were assumed to arise by capture of the configurationally stable ion pair, R^+X^-.

$$\mathrm{RX} \underset{k_{-1}}{\overset{k_1}{\rightleftharpoons}} \mathrm{R^+X^-} \begin{cases} \xrightarrow{k_s} \mathrm{ROS} \\ \xrightarrow{k_N[N_3^-]} \mathrm{RN_3} \end{cases} \quad [4.19]$$

By applying the steady state approximation to [*4.19*] we may obtain (4.8) for the observed first-order rate constant, k_{obs}, for the consumption of the substrate RX. The first-order rate constant for solvolysis, k_{solv},

$$k_{obs} = k_1(k_s + k_N[N_3^-])/(k_{-1} + k_s + k_N[N_3^-]) \qquad (4.8)$$

that would be observed in the presence of an inert salt at the same concentration as the sodium azide in [*4.19*] is given by (4.9). From (4.8) and

$$k_{solv} = k_1 k_s/(k_{-1} + k_s) \qquad (4.9)$$

(4.9) may be obtained (4.10), in which the substitutions $m = k_N/k_s$ and $x = k_{-1}/k_s$ have been made. If x is treated as an adjustable parameter (4.10) may be evaluated from the product composition, making use of

$$\frac{k_{obs}}{k_{solv}} = \frac{(x + 1)(1 + m[N_3^-])}{(x + 1 + m[N_3^-])} \qquad (4.10)$$

the relationship (4.11). The ratio k_{obs}/k_{solv} may also be evaluated from

$$\%RN_3/\%ROS = k_N[N_3^-]/k_s = m[N_3^-] \qquad (4.11)$$

the kinetic results by assuming that the added sodium azide produces only a normal salt effect on the rate of hydrolysis (4.12). The value of

$$k_{solv} = k_{solv}^0(1 + b[N_3^-]) \qquad (4.12)$$

k_{solv}^0 in this equation is given by the observed rate constant, k_{obs}, in the absence of salt. The required ratio may therefore be calculated from (4.13).

$$k_{obs}/k_{solv} = k_{obs}/k_{solv}^0(1 + b[NaN_3]) \qquad (4.13)$$

The experimental results for the hydrolysis of 2-octyl methanesulphonate in 30 per cent aqueous dioxane are summarised in table 4.6. In order to obtain values of k_{obs}/k_{solv} from the observed rate constants, k_{obs}, it was assumed that sodium azide would have the same normal salt effect on the solvolysis as that produced by a weakly nucleophilic salt. The kinetic ratios were therefore calculated using (4.13) with $b = 1.04$, the value determined for lithium perchlorate; these are shown as the open circles in fig. 4.3. The smooth curve drawn through the circles (II) was calculated from the observed product compositions using (4.10) and assuming $x = 2.59$. The agreement found was assumed to be good evidence for the ion-pair scheme [*4.19*]. The straight line (I), also included in fig. 4.3, is the variation of k_{obs}/k_{solv} calculated for the S_N2 scheme

TABLE 4.6 *The effect of sodium azide on the hydrolysis of 2-octyl methanesulphonate in 30 per cent aqueous dioxane at 36.2 °C*[a]

$[NaN_3]$ (mol l^{-1})	% RN_3	$10^4 k_{obs}$ (s^{-1})
0	0	1.74 ± 0.04
0.0543	38.4	2.43 ± 0.08
0.0571	39.3	2.26 ± 0.11
0.0979	52.5	2.67 ± 0.14
0.152	54.5	3.65 ± 0.14
0.199	64.0	3.73 ± 0.12
0.258	69.3	4.71 ± 0.08
0.311	74.6	4.91 ± 0.15

[a] Data from Sneen and Larsen (1969).

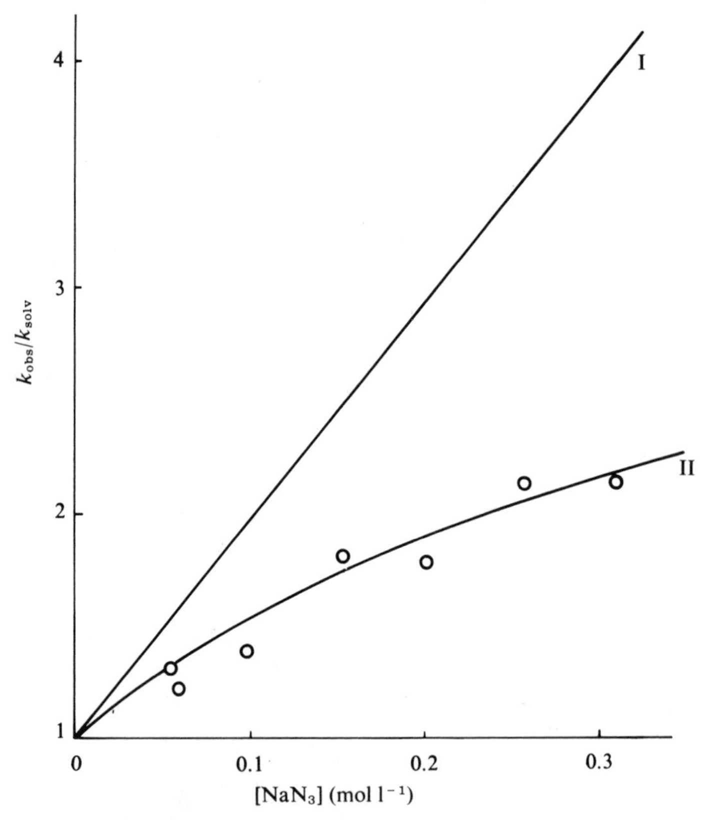

Fig. 4.3 The effect of sodium azide on the ratio k_{obs}/k_{solv}. Solid lines are calculated from product compositions for: I, S_N2 mechanism; II, ion-pair mechanism. Open circles are ratios calculated from kinetic data with $b = 1.04$.

[*4.20*]. It can be shown that for this scheme the required ratio is given by (4.14), i.e. a linear relationship is to be expected between k_{obs}/k_{solv} and the concentration of sodium azide.

$$RX \xrightarrow{k_s} ROS$$
$$RX \xrightarrow{k_N[N_3^-]} RN_3 \qquad [4.20]$$

$$k_{obs}/k_{solv} = 1 + m[N_3^-] \qquad (4.14)$$

It has been pointed out, however, that the type of correlation that is observed between the kinetic and product data is dependent upon the assumptions made about the nature of salt effects on the solvolysis (Raber, Harris, Hall and Schleyer, 1971). In particular, since salt effects can be specific (§3.2), it may not be valid to use the value of b determined for lithium perchlorate to calculate the effect of sodium azide on the rate of solvolysis. If the assumption is made that $b = -1.0$, the kinetic ratios calculated by (4.13) in fact agree very well with the predictions of the S_N2 mechanism (fig. 4.4). A negative salt effect might not be unreasonable for the reaction in question.

Because of the uncertainty in correcting for the effect of added salt on the rate of solvolysis, the results obtained with sodium azide do not provide unequivocal support for the ion-pair scheme [*4.19*]. Moreover, they need not support the suggestion that this scheme forms the basis of a 'unified' mechanism of nucleophilic substitution (Sneen and Larsen, 1969). According to this hypothesis, nucleophilic substitution always involves an ion pair, the formation of which is rate limiting at one extreme (corresponding to the S_N1 mechanism), and the capture of which by a nucleophile is rate limiting at the other extreme (corresponding to the S_N2 mechanism). Borderline reactions then correspond to those cases in which formation of the ion pair and its capture are competitive processes. An important feature of the unified mechanism is that a direct nucleophilic attack on the covalent starting material is excluded. If this hypothesis is accepted, and if only a single ion-pair intermediate is considered to be involved (as required by [*4.19*]), then it becomes difficult to explain the effects of added salts on the decompositions of *p*-methoxybenzyl and *p*-phenoxybenzyl chlorides in 70 per cent aqueous acetone (Gregory, Kohnstam, Queen and Reid, 1971).

In this solvent both the benzyl chlorides are believed to undergo solvolysis by a unimolecular mechanism. In terms of mechanism [*4.19*] this means that k_1 will be rate limiting; added salt will compete with the solvent for the ion pair formed, and, if a direct displacement on covalent material is excluded, the only effect on the rate of decomposition of RX will be a normal salt effect on k_1. The results (table 4.7) show that with

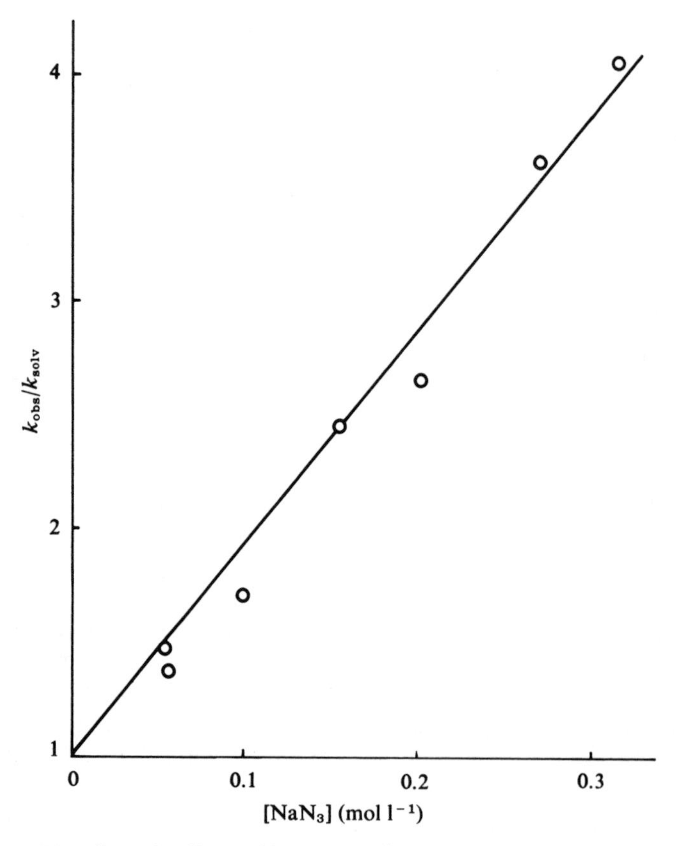

Fig. 4.4 The effect of sodium azide on the ratio k_{obs}/k_{solv}. Solid line is calculated from product compositions for S_N2 mechanism. Open circles are ratios calculated from kinetic data with $b = -1.0$.

benzhydryl chloride only normal salt effects are observed. With the two benzyl compounds, on the other hand, although normal salt effects are observed with the weakly nucleophilic salts, rate enhancements that cannot reasonably be attributed to normal salt effects are observed with the more nucleophilic salts sodium bromide and sodium azide. The last-mentioned results can only be accommodated by scheme [*4.19*] if a direct displacement on covalent RX, i.e. an S_N2 pathway, is allowed.

TABLE 4.7 *The effect of added salts on the rates of decomposition of benzhydryl chloride, p-methoxybenzyl chloride, and p-phenoxybenzyl chloride in 70 per cent aqueous acetone at 20 °C*[a]

Salt	$(C_6H_5)_2CHCl$		p-$CH_3O.C_6H_4.CH_2Cl$		p-$C_6H_5O.C_6H_4.CH_2Cl$	
	$k_t/k_t^{o\,b}$	b^c	$k_t/k_t^{o\,b}$	b^c	$k_t/k_t^{o\,b}$	b^c
$NaClO_4$	1.151	3.02	1.159	3.18	1.128	2.56
$NaBF_4$	1.120	2.40	1.131	2.62	1.110	2.20
$NaNO_3$	1.070	1.40	1.144	2.88	1.109	2.18
NaBr	1.091	1.82	1.255	5.10	2.49	29.8
NaN_3	1.136	2.72	1.735	14.7	20.0	380

[a] Data from Gregory, Kohnstam, Queen and Reid (1971).
[b] All salt effects measured at a salt concentration of 0.05 mol l^{-1}.
[c] Apparent values of the normal salt parameter, calculated from the given values of k_t/k_t^o.

Alternatively, the results can be explained in terms of an ion-pair scheme which involves more than one type of ion pair, in which case the S_N2 pathway is not required. In any event a mechanism which involves only a single pathway to the reaction products seems to be excluded.

4.3.5. Primary systems. The reactions of simple primary alkyl substrates in the common solvents suggest that these systems usually undergo solvolysis by a single mechanism which has all the defining characteristics of the S_N2 mechanism. It must be admitted, however, that the suggestion (Sneen and Larsen, 1969) that processes which are called S_N2 may in fact involve a rate-limiting attack of solvent (or other nucleophile) on a preformed ion pair also remains a possibility. Distinguishing unambiguously between the two mechanisms may prove to be extremely difficult.

In recent years trifluoroethanol (TFE) and trifluoroacetic acid (TFA) have been added to the range of solvents which are used to study solvolysis reactions. Both of the solvents are characterised by being weakly nucleophilic and having high ionising power. Thus isopropyl brosylate, which in other solvents displays either S_N2 or borderline behaviour, in both TFE and TFA shows typical S_N1 behaviour. In contrast with this result, ethyl trifluoromethanesulphonate reacts even in TFA by a mechanism which is far from being S_N1 (Dafforn and Streitwieser,

1970). The secondary α-deuterium isotope effect, for example, was found to be 1.063 (per α-D) at 50 °C, which is much smaller than the maximum value of 1.23 expected for an S_N1 reaction of a sulphonate ester. This result indicates that the solvolysis of ethyl sulphonates in the other common solvents, all much more nucleophilic than TFA, will occur predominantly by the S_N2 mechanism.

The order of reactivity for the series of tosylates Me, Et, n-Pr, i-Bu and neopentyl changes from the steeply descending one in ethanol, expected for a series of S_N2 reactions, to the steeply ascending one in TFA (table 4.8). The order of reactivity observed in TFA suggests that a

TABLE 4.8 *The relative rates of solvolysis of some primary alkyl tosylates*[a]

	Me	Et	n-Pr	i-Bu	neopentyl
EtOH	4000	1750	–	80	1
TFA	1	13	93	3060	5970

[a] Data from Reich, Diaz and Winstein (1969).

mechanism involving ionisation has assumed importance. Product studies show that whereas MeOTs and EtOTs give the corresponding trifluoroacetates in quantitative yield, n-PrOTs gives predominantly, and i-BuOTs and neopentylOTs give totally, rearranged triuoroacetates. The large rate enhancements and the formation of products with rearranged structures in TFA are most easily explained in terms of anchimeric assistance to ionisation (§5.2.2). The results thus indicate that a limiting solvolysis with concerted migration of a neighbouring group occurs (90) and (91), rather than a process which involves the formation

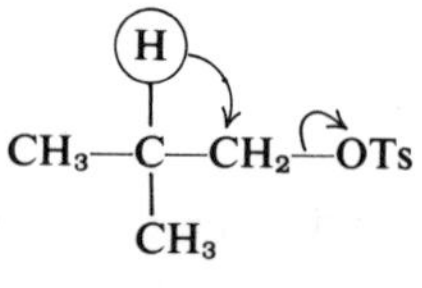

$CH_3—C(H)(CH_3)—CH_2—OTs$

(90)

$CH_3—C(CH_3)(CH_3)—CH_2—OTs$

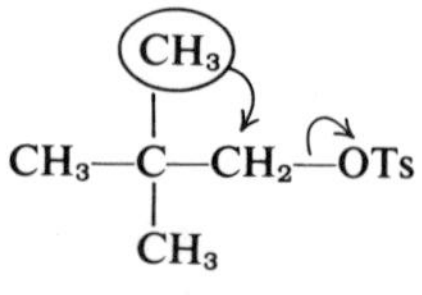

(91)

of a primary alkyl carbonium ion. These observations therefore provide further confirmation that the normal reactions of primary alkyl systems involve nucleophilic participation, by either an external or an internal nucleophile, and that primary carbonium ions are apparently never involved.

4.3.6. Summary and additional comments. Some of the experimental evidence upon which our current views of nucleophilic substitutions are based has been presented in the preceding sections of this chapter; in this last section we shall summarise this evidence and discuss the present position.

The basic assumption, that of ionic association, is not in question, and the structural interpretation of ionic interactions in terms of ion pairs has become well established. It is usual in discussions of the mechanisms of nucleophilic substitutions to use the terminology proposed by Winstein, and so for many reactions it is necessary to include two intermediates, the intimate and the solvent separated ion pair, in addition to free carbonium ions. The structures of the two types of ion pair are considered to be different, as implied by their names, and indeed there is some independent evidence that ion pairs having different discrete structures do exist (Szwarc, 1969). However, little is known about the nature of the intermediates involved in nucleophilic substitutions. This in no way detracts from the usefulness of the hypothesis including ion pairs, for it is sufficient merely to regard the intermediates as ionised species in which the counter ions behave as a kinetic entity.

Since the structures of the intermediates are unknown there is no reason to prefer the description in terms of ion pairs over that in terms of other ionic aggregates in many cases, and a scheme involving ion quadrupoles has been suggested (Ingold, 1969). Nevertheless, the ion pair appears to be preferred and has been adopted by common usage, and most discussions of mechanism centre on the dual ion-pair scheme of Winstein [*4.21*] (cf. Winstein, Appel, Baker and Diaz, 1965).

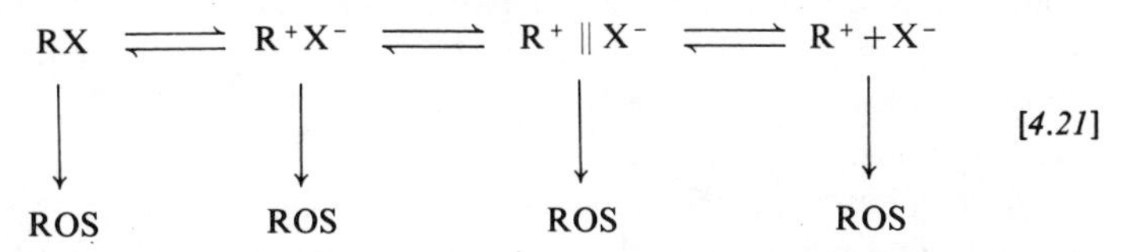

[*4.21*]

In this scheme the ion pairs represent various stages in the ionisation and dissociation of the covalent substrate to the fully dissociated (i.e. kinetically free) carbonium ion, and in principle reaction with a nucleophile may occur at any of the stages, as indicated. The number of kinetically distinct species required in any given reaction will depend upon the solvent and the nature of the substrate. Originally scheme [*4.21*] was proposed to explain the acetolysis reactions of certain substrates capable of forming relatively stable carbonium ions, but now its use has been extended to include other solvents and substrates of many structural types.

In general the reactions of tertiary alkyl compounds and secondary alkyl compounds that give rise to stable carbonium ions, i.e. reactions which are usually classified as limiting (S_N1), are now believed to require at least one type of ion-pair intermediate, even apparently in quite polar solvents; examples are mentioned in §4.3.2 and §4.3.3. The rate-limiting step for such reactions is either the initial ionisation of the substrate or the subsequent diffusion apart of the counter ions, as in the conversion of the intimate to the solvent separated ion pair. The nucleophile is assumed not to be involved in this step and the product arises by capture of one of the carbonium-ion species in a subsequent step.

More recently the ion-pair hypothesis has been used to describe the reactions of simple secondary alkyl compounds (Sneen and Larsen, 1969) and of primary alkyl compounds (Scott, 1970), i.e. reactions which are usually classified as borderline nucleophilic or S_N2 processes. Such reactions are characterised by being dependent upon the nature of the nucleophile, but whether or not this necessarily implies the presence of the nucleophile in the transition state of the rate-limiting step for all borderline reactions will be considered in the next chapter (§5.5). If, for the present, we assume that for all reactions in the nucleophilic category the nucleophile is present in the rate-limiting step, and if we further assume that the formation of an ion pair, scheme [*4.21*]. may not always be rate limiting, then we might expect that product could arise by the rate-limiting attack of the nucleophile either on the covalent substrate or on an ion-pair intermediate. The latter possibility affords an attractive explanation of borderline behaviour, since a direct nucleophilic displacement on a partly ionised species might be expected to display characteristics intermediate between those of a limiting (S_N1) reaction and a classical S_N2 reaction.

There is some experimental evidence to suggest that the solvolysis reactions of 2-octyl arenesulphonates proceed, in part at least, by a pathway which involves an ion-pair intermediate (§4.3.4). In particular the ^{18}O-equilibration experiments of Diaz, Lazdins and Winstein (1968*a*) demonstrate clearly that internal return occurs, which is taken to indicate that a solvolysis pathway involving an ion-pair intermediate is required. However, the results give no indication of the importance of that pathway in the overall reaction; it is possible that the major part of the product could arise from a direct displacement on the covalent substrate.

A similar conclusion is to be drawn from the results obtained in a study of the solvolysis of 2-octyl methanesulphonate in aqueous dioxane in the presence of azide ion (Sneen and Larsen, 1969). Sneen and Larsen interpret the results in terms of an ion-pair mechanism, but, as pointed out in §4.3.4, the uncertainty about the nature of the salt effect produced by sodium azide allows no firm conclusions to be drawn from their results, which could be consistent with a classical S_N2 mechanism. At present therefore the mechanisms of nucleophilic substitutions involving simple secondary systems remain in question and there seems to be no indication yet as to whether these reactions proceed by a single mechanism or by a mixture of mechanisms, although the evidence does seem to suggest that ion-pair intermediates may be involved.

The significance of the work of Sneen and Larsen lies not only in their extension of the ion-pair scheme [*4.21*] to include borderline reactions, but also in their assertion that nucleophilic displacements always occur on an ion pair and that attack on the covalent substrate is unimportant. In their view a reaction which could be classified as S_N2 is to be described as a rate-limiting attack of the nucleophile on a preformed (intimate) ion pair. This represents a departure from the traditional view of the bimolecular mechanism for nucleophilic substitution which assumes that it is a one-step concerted process. The mechanistic distinction to be made between limiting (S_N1) and nucleophilic (S_N2) reactions therefore becomes more difficult, and indeed Sneen and Larsen describe the ion-pair scheme as providing a unified mechanism for nucleophilic substitutions.

Although the idea of a single mechanism may appear attractive, there is no compelling reason for it to be adopted at present; as pointed out in §4.3.4 the experimental evidence used to support it is not unequivocal. Scott (1970) has analysed the reactions of the methyl halides with various nucleophiles in water and has concluded that the behaviour is consistent with the scheme of Sneen and Larsen, but clearly more examples are required before this scheme can be accepted as being firmly established. For the present, the nucleophilic reactions of primary systems are still best regarded as being classical S_N2 reactions.

In recent years the basis of the duality of mechanism has been questioned, but there appears to be no need to abandon it at present, although some modification to the traditional views of the mechanisms of nucleophilic substitutions is necessary (cf. Raber and Harris, 1972).

5 Intramolecular interactions

5.1. Introduction

The discussion in chapter 2 of the effects of substituents on the reactivity of a molecule towards nucleophilic substitution was restricted to a consideration of electronic and steric effects. In this chapter we shall consider the important effect of neighbouring group participation, i.e. a direct, intramolecular, interaction between a substituent and the reaction centre. Such an interaction may influence the reaction, either by stabilising the transition state for substitution by an external nucleophile, or by the substituent becoming permanently bonded to the reaction centre. Many examples of neighbouring group participation occurring in nucleophilic displacements at saturated carbon centres have been described (Capon, 1964).

5.1.1. Some different types of interactions. Several examples of neighbouring group participation have already been mentioned in earlier chapters. A substituent may interact with the reaction centre if it contains an atom with lone pairs of electrons, for example the α-carboxylate anion in the alkaline hydrolysis of the 2-bromopropionate ion [*5.1*],

[*5.1*]

and the ω-methoxy group in the acetolysis of 4-methoxybutyl brosylate [*5.2*]. Neighbouring group participation is more evident in the alkaline

[*5.2*]

hydrolysis of 4-chlorobutanol, because tetrahydrofuran may be isolated from the reaction products [*5.3*].

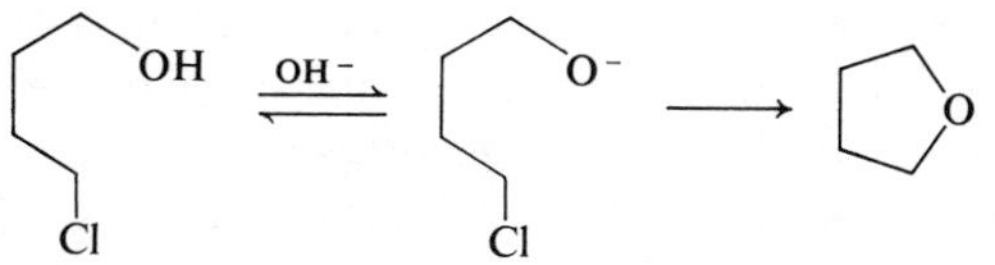

[*5.3*]

Neighbouring phenyl groups may interact with the reaction centre, and 'phenonium' ions are frequently assumed to be involved as intermediates in solvolysis reactions. The acetolysis of neophyl brosylate [*5.4*] leads to the rearranged acetate and the bridged carbonium ion is thought to be the product of the initial ionisation.

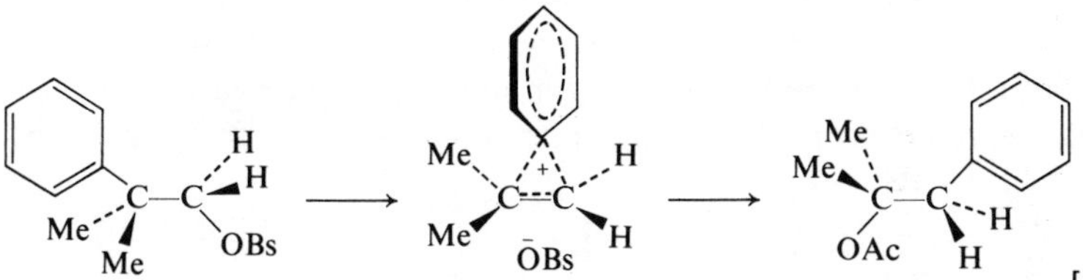

[*5.4*]

A saturated carbon atom may act as the neighbouring group, in which case the electrons of a σ-bond interact with the reaction centre [*5.5*]. This type of bridged structure is frequently referred to as a 'non-classical' carbonium ion; similar structures are probably involved in

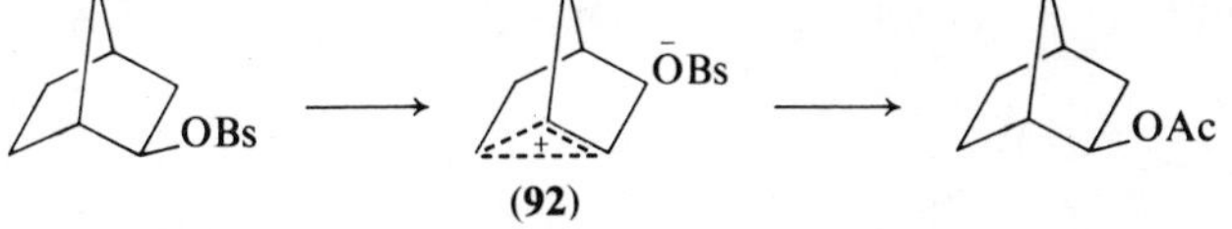

(92) [*5.5*]

many Wagner–Meerwein rearrangements. The norbornyl cation (92) is also thought to be involved as an intermediate in the acetolysis of 2-(cyclopent-3-enyl)ethyl *p*-nitrobenzenesulphonate (93) to give *exo*-2-norbornyl acetate (94) [*5.6*]. In this example the electrons of the π-bond

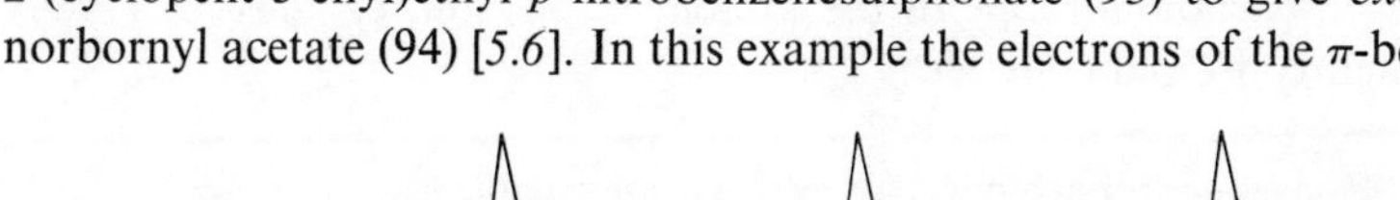

p-$NO_2.C_6H_4.SO_2.O$ **(93)** **(94)**

[*5.6*]

interact with the reaction centre; this represents the π-route to the norbornyl cation, whereas the previous example represents the σ-route.

Bridged carbonium ions may be classified according to whether or not they contain a sufficient number of electrons to allow a bridged representation of the ion to be drawn with normal single bonds. The phenonium ion shown in [*5.4*] is an example of an electron-sufficient system, because it can be represented by the canonical structures [*5.7*] in each of

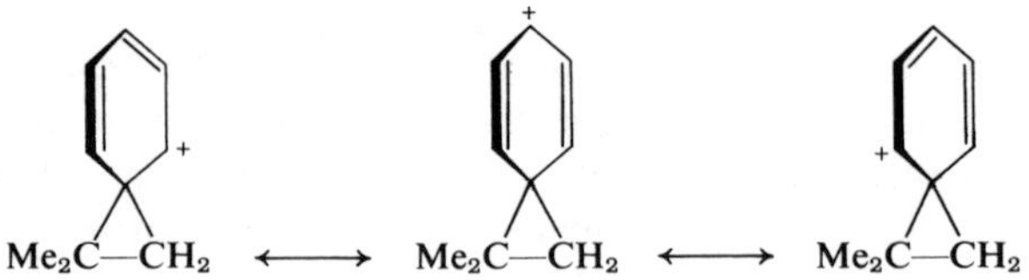

Me_2C—CH_2 ⟷ Me_2C—CH_2 ⟷ Me_2C—CH_2 [*5.7*]

which it is possible to draw a cyclopropane ring with normal single bonds. It is possible that no-bond structures such as (95) also contribute to the stability of the bridged system, and this is implied in the more usual representation of a phenonium ion [*5.4*]. The bridging in the

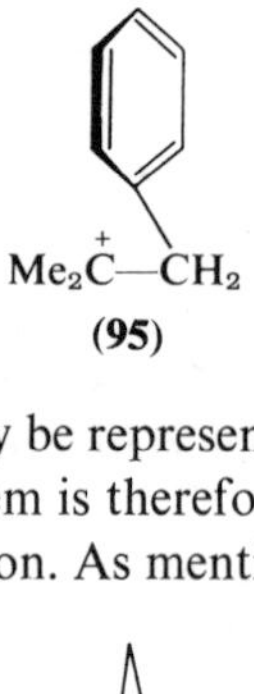

$Me_2\overset{+}{C}$—CH_2

(95)

norbornyl cation (92) can only be represented by the no-bond canonical structures [*5.8*], and this system is therefore an example of an electron-deficient bridged carbonium ion. As mentioned above, such systems are

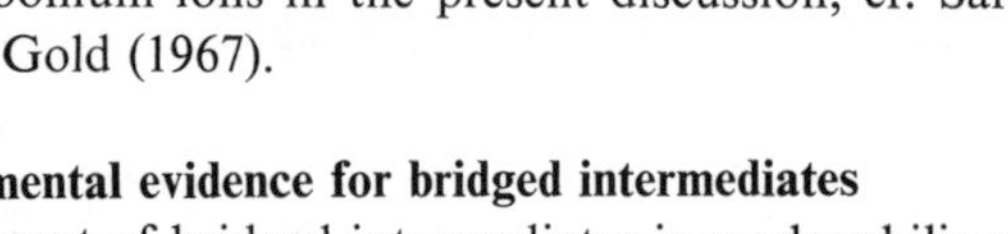

⟷ ⟷ [*5.8*]

usually called non-classical, but they will be referred to merely as bridged carbonium ions in the present discussion, cf. Sargent (1966), Bethell and Gold (1967).

5.2. Experimental evidence for bridged intermediates

The involvement of bridged intermediates in nucleophilic substitutions has been demonstrated unequivocally in a few favourable cases by the isolation of the bridged species. In most cases, however, the involvement of such a species in a reaction pathway has been inferred from other experimental evidence. This has most often come from the observation of unexpected products, and the observation of unexpectedly high rates of reaction. Additional evidence has also become available from measurements of secondary deuterium isotope effects, and from the application of physical methods, such as n.m.r. spectroscopy.

5.2.1. Product studies. One indication that participation by a neighbouring group has occurred during a nucleophilic substitution is the formation of products in which the bridging group has migrated, or in which the ring structure is retained. For substitution at an asymmetric carbon atom, the formation of products with retention of configuration is also good evidence for the intervention of a bridged intermediate. The solvolysis of the 2-bromopropionate ion [*5.1*], for example, gives product with complete retention of configuration (§3.3.4).

Acetolysis of both *endo*- and *exo*-norborn-5-en-2-yl brosylates (96) and (99), gives *exo*-norborn-5-en-2-yl acetate (100) and 3-acetoxytricyclo[2,2,1,0]heptane (101), [*5.9*]. The formation of only the *exo*-acetate

OBs (96) → + (97) → + (98) ← OBs (99) [*5.9*]

(98) → OAc (100); AcO (101)

(100), and the bridged product (101), is most easily accounted for by assuming that both arise from the bridged intermediate (98); the bridging prevents *endo*-attack of acetate and so accounts for the specific nature of the products. The *exo*-isomer (99) reacts about 8000 times more rapidly than the *endo*-isomer (96), which is taken to indicate that the ionisation of (99) is assisted by an interaction involving the π-bond leading to the bridged carbonium ion (98). Such an interaction is not possible with (96), owing to the presence of the departing anion, and so ionisation gives first the classical carbonium ion (97), which then rearranges to the more stable bridged carbonium ion. The increased rate of reaction of (99), that arises because of the involvement of a more stable bridged transition state in the ionisation, is termed anchimeric acceleration (§5.2.2).

In many nucleophilic substitutions a rearrangement, involving migration of the neighbouring group, accompanies the substitution, as in the acetolysis of neophyl brosylate [*5.4*]. The observation of a rearranged product does not require the intermediacy of a bridged structure, since it might be simply a consequence of the conversion of the first formed open carbonium ion into another more stable open carbonium ion (cf. fig. 5.1). However, if the rearrangement is accompanied by an anchimerically accelerated rate of reaction then it does indicate that bridging is important in the transition state of the rate-limiting step.

The compounds 4-methoxypentyl brosylate (102) and 4-methoxy-l-methylbutyl brosylate (103) undergo acetolysis at about the same rate and give the same mixture of products (105) 60 per cent, and (106) 40 per cent. The identical product compositions indicate that a common intermediate is involved in both reactions, and this is most reasonably assumed to be the cyclic oxonium ion (104).

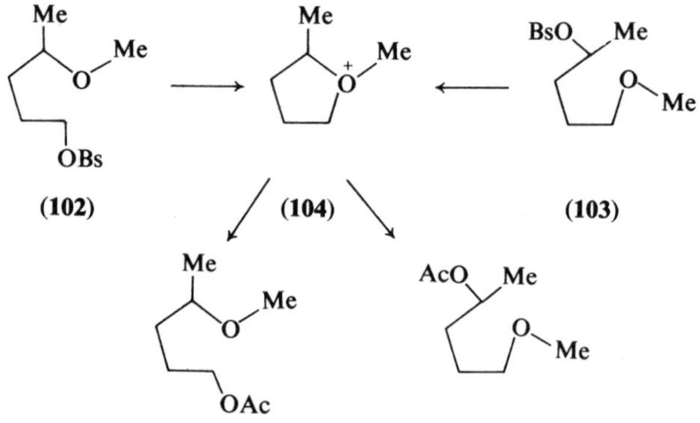

Me Me O OBs **(102)** Me Me O+ **(104)** BsO Me O Me **(103)**

Me Me O OAc **(105)** AcO Me O Me **(106)**

5.2.2. Kinetic studies. A compound capable of forming a bridged carbonium ion has two potential reaction pathways available for a unimolecular nucleophilic substitution [*5.10*], the anchimerically assisted pathway, k_Δ, and the anchimerically unassisted pathway, k_s. A reaction will proceed by both pathways if the transition states leading to the different carbonium ions are of similar energy, i.e. if both pathways have similar Gibbs free energies of activation. This situation might arise if the participation by the neighbouring group does not significantly stabilise the transition state leading to the bridged ion, or if formation of the open

k_s → product

k_Δ → product

[5.10]

ion is itself an assisted process, possibly involving nucleophilic participation by the solvent (§5.5). In many cases, however, the transition state leading to the bridged ion has the lower energy (fig. 5.1) and the anchimerically assisted mechanism is the favoured reaction pathway.

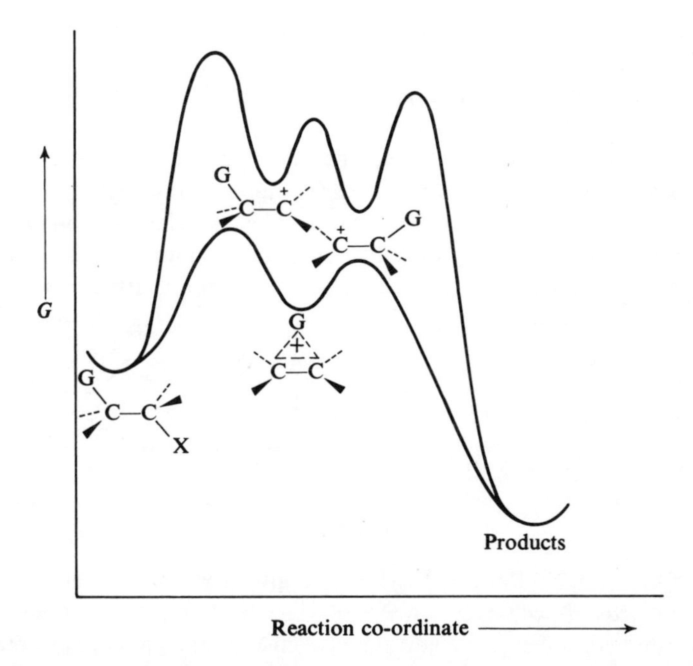

Fig. 5.1 Rate profiles illustrating the changes in free energy for the anchimerically assisted and unassisted pathways.

When the formation of a bridged ion is possible, therefore, the rate of a substitution reaction may be much faster (due to anchimeric acceleration) than that expected for the same process involving the unbridged ion. Rate enhancements in the range 10^2 to 10^4 are frequently observed, and used as evidence for the formation of a bridged intermediate, or, more correctly, as evidence for bridging in the transition state. However, much smaller values can still be consistent with a mechanism in which the major part of the product is formed via a bridged ion. For scheme [*5.10*], the proportion of product formed by the bridged ion (k_Δ/k_t) is related to the observed rate enhancement (k_t/k_s) by (5.1). In

$$k_\Delta/k_t = (1 - k_s/k_t) \qquad (5.1)$$

this equation $k_t = (k_s + k_\Delta)$ and is the observed rate constant for reaction. Thus 75 per cent of the product will be formed via the bridged ion for an observed rate enhancement of only 4.

Regardless of the magnitude of the anchimeric acceleration, one problem always remains when a rate enhancement is used as a criterion of bridging, that of estimating the rate to be expected in the absence of bridging. Various methods have been used, but because of the uncertainties involved they are not reliable for detecting enhancements of less than an order of magnitude.

A suitable model compound may be used to give the unassisted rate; this may be either a stereoisomer of the compound in question, but one for which bridging is impossible; or it may be a similar compound that does not contain the bridging substituent. For example, *cis*-2-acetoxycyclohexyl tosylate (107) undergoes acetolysis about 10^4 times more

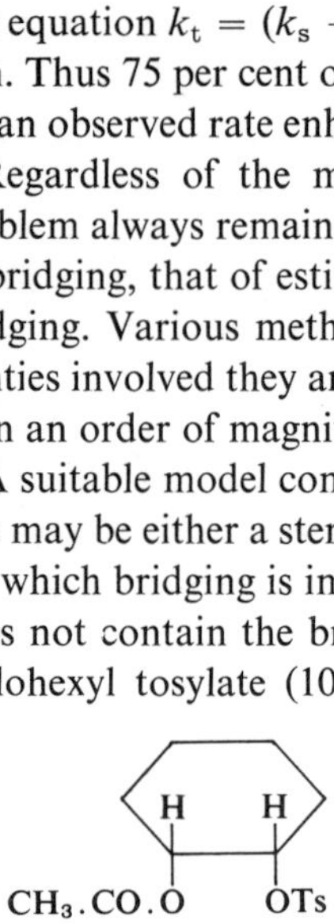

H H
CH_3.CO.O OTs
(107)

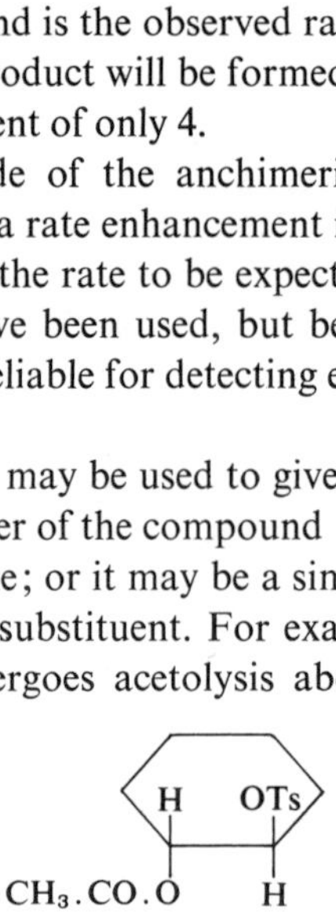

H OTs
CH_3.CO.O H
(108)

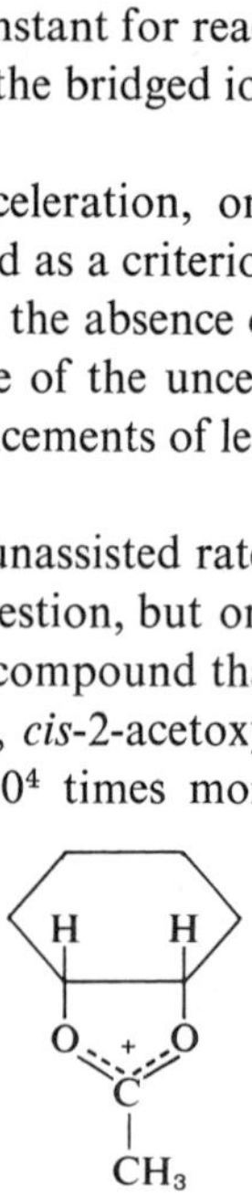

H H
O + O
C
CH_3
(109)

slowly than does cyclohexyl tosylate and gives the *trans*-diacetate. The isomeric *trans*-2-acetoxycyclohexyl tosylate (108) reacts at about the same rate as cyclohexyl tosylate and also gives, in the presence of potassium acetate, the *trans*-diacetate. The enhanced rate of reaction of the *trans*-isomer can be explained in terms of an assisted ionisation to give the bridged intermediate (109), which also accounts for the formation of the *trans*-product; similar assistance is not possible in the case of the *cis*-isomer.

Stereochemical studies suggest that a phenonium ion is an intermediate in the acetolysis of 3-phenyl-2-butyl tosylate (110). A suitable reference compound for estimating the anchimeric acceleration would appear to be 2-butyl tosylate (111), but the observed rate constants for acetolysis show that (110) reacts more slowly than (111). When 'corrections' are applied to take into account the rate retarding inductive effect of the phenyl group, the steric inhibition to solvation caused by

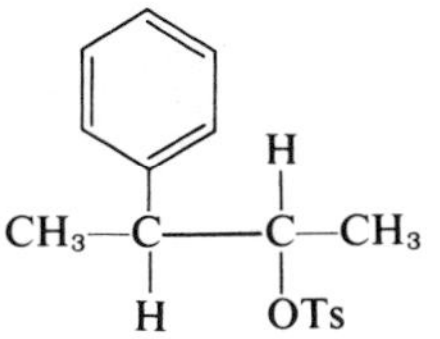

$$CH_3-CH(C_6H_5)-CH(OTs)-CH_3 \qquad CH_3-CH_2-CH(OTs)-CH_3$$

(110) 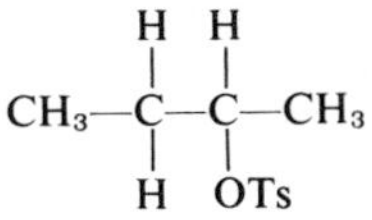 **(111)**

the phenyl group, and the effect of internal return (see p. 159), the order of reactivity is reversed. The actual rate enhancement calculated for (110) depends upon the assumptions used in making the various corrections, and values between 3 and 70 may be obtained.

Linear free energy relationships have also been used to estimate unassisted rates of nucleophilic substitutions, and an example of this was noted in an earlier section (§3.6.2). The disadvantage of this approach is that kinetic data for several structurally related compounds are required to establish a correlation line for the unassisted reaction. Nevertheless, this method has been usefully applied to several series of compounds of the 3-aryl-2-butyl type of structure in an attempt to estimate the importance of the assisted pathway (k_Δ) in systems reacting by competing pathways (§5.4.2).

Estimates of enhanced rates have also been made by using calculated values for the unassisted rate constants, one of the most successful attempts being the semi-empirical approach of Schleyer (1964) (§2.1.4). In this method estimates are made for changes in bond angle strain, torsional strain, non-bonding interactions, and polar effects on going from the ground state to the transition state for ionisation. Some examples of rate constants calculated by this procedure are given in table 5.1. For compounds that are thought to react via bridged intermediates, the calculated rate constant is much smaller than the observed value, and the ratio of the observed to the calculated rate constant is taken as a measure of the anchimeric acceleration.

Care should always be taken when using model compounds to provide a measure of anchimeric acceleration, because there are several other factors which may produce an enhanced rate of reaction. In particular, it is necessary to ensure that the compounds being compared react by the same mechanism, and that the model compound does not react unusually slowly for some special structural reason. In some cases the possibility may have to be considered that the accelerated rate is due to the release of steric strain accompanying ionisation. A more serious difficulty arises when observed titrimetric rate constants are used to

TABLE 5.1 *Calculated rate constants for the S_N1 acetolysis of some secondary alkyl tosylates*[a]

	$\log k_{rel}$		$\log(k_{rel}^{obs}/k_{rel}^{calc})$
	Calc	Obs	
(*a*) Unassisted			
Cyclohexyl	−0.1	(0.00)	0.1
7-Norbornyl	−7.0	−7.00	0.0
endo-2-Norbornyl	−0.2	+0.18	0.4
endo-2-Norbornenyl	−1.0	−1.48	−0.5
2-Butyl	−0.3	+0.53	0.8
3-Methyl-2-butyl	+0.6	+0.93	0.3
3,3-Dimethyl-2-butyl	+1.5	+0.62	−0.9
(*b*) Anchimerically assisted			
anti-7-Norbornenyl	−8.8	4.11	12.9
syn-7-Norbornenyl	−8.9	−3.28	5.6
Cyclobutyl	−4.2	0.99	5.2
exo-2-Norbornenyl	−1.4	2.42	3.8
exo-2-Norbornyl	−0.6	2.71	3.3

[a] Data from Schleyer (1964); relative to cyclohexyl tosylate.

calculate anchimeric accelerations, because k_t may be a composite term involving constants for ionisation, dissociation and ion-pair return. An increase in k_t is usually assumed to indicate an increase in the rate of ionisation, but it may arise from a decrease in the amount of ion-pair return. An enhanced rate of reaction is only evidence for anchimeric acceleration if it is associated with an increase in the rate of ionisation.

5.2.3. Secondary deuterium isotope effects. The observation of anchimeric acceleration implies an interaction between a neighbouring group and the reaction centre in the transition state of the rate-limiting step. If the proximity of the neighbouring group hinders the bending motions of a hydrogen atom attached to the reaction centre, the secondary α-D effect for the reaction will be less than the value expected for the reaction in the absence of bridging (§3.4.1). The magnitude of the α-D effect may therefore provide information about the nature of the participation in the transition state.

The ethanolysis of compounds (102) and (103) is assumed to proceed (as in the case of acetolysis) through the same cyclic oxonium ion intermediate (104). The α-D effect has been measured for the solvolysis of each compound in 96 per cent aqueous ethanol at 40.5 °C (Sunko and Borčić, 1970). Whereas the primary compound (102) gives an α-effect of unity, $k_H/k_D = 1.00$, the secondary compound (103) shows a small normal α-effect, $k_H/k_D = 1.08$. These results suggest that the transition state for the formation of (104) from the primary compound (112) has a structure which resembles that for an S_N2 process, and that the transition state from the secondary compound (113) has more ion-pair character (or at least a structure which permits greater vibrational freedom for the

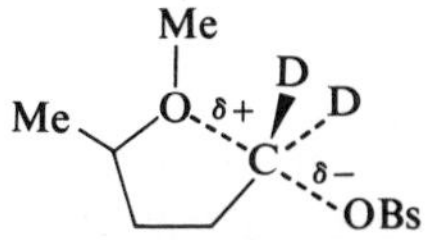

Me
Me O δ+ D D
C δ−
OBs
(112)

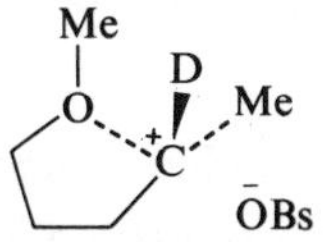

Me
O D Me
+ C
OBs
(113)

α-hydrogen atom). The assumption that in this example the larger isotope effect indicates more carbonium-ion character in the transition state, is consistent with the known behaviour of simple primary and secondary systems in solvolysis (§4.3).

The 2-arylethyl arenesulphonates (114*a*, 114*b*, and 114*c*) all undergo formolysis by the anchimerically assisted pathway. The observed α-D isotope effects (table 5.2) are considerably smaller than the expected maximum effect of 1.23 (§3.4.2), consistent with bridging in the transition state (Lee and Noszkó, 1966). The α-D isotope effects reported by Lee and Noszkó show a trend towards higher values as the 2-aryl group changes from phenyl to 2,4-dimethoxyphenyl, which these authors rationalise in terms of Hammond's hypothesis. As applied to reactions involving bridged ions, this predicts that as the bridged ion becomes more stable so the bridging is less developed in the transition state (115). However, it can be argued that if less bridging in the transition state implies a structure which is more reactant-like, then there should also be a stronger interaction between the leaving group and the reaction centre, so that the overall effect on the magnitude of the α-D isotope effect is difficult to predict. Moreover, the trend discussed by Lee and Noszkó is not really apparent in the data, for when the observed isotope effects are

TABLE 5.2 *Secondary α-deuterium isotope effects for formolysis of 2-arylethyl compounds*

		R_1	R_2	X	k_H/k_D obs[a]	k_H/k_D corr[b]
(114): R_1-, R_2-substituted phenyl–H_2C–CD_2X	*a*	H	H	OTs	1.09	1.113
	b	OMe	H	OTs	1.10	1.105
	c	OMe	OMe	OBs	1.12	1.113

[a] Data from Lee and Noszkó (1966).
[b] Reported values of k_H/k_D (per α-D) corrected for deuterium content and extrapolated to 25 °C.

corrected for the different deuterium contents of the deuterated materials, and the results are extrapolated to the same temperature, there is no significant difference in the values.

(115)

The reason for the near constancy of the α-D effect is not clear. It may indicate a fortuitous cancellation of effects, as discussed above, or it may indicate that the magnitude of the α-D effect is not particularly sensitive to a variation in the amount of bridging in this particular series of compounds. The fact that the α-D effect is less than the maximum value might then simply be an indication that the rate-limiting step is the formation of an intimate ion pair (§4.2.5). Minor changes in the degree of bonding (to both incoming and leaving groups) in the transition state need not affect the magnitude of the α-D effect appreciably if the two are related as shown in fig. 5.2. The value of k_H/k_D might be sensitive to changes in the degree of bonding only when bonding is relatively important; beyond a certain degree of bonding there might be little change in the value of k_H/k_D. The preceding discussion is sufficient to show that there is, in general, no simple connection between the magnitude of the α-D isotope effect and the amount of bridging in the transition state.

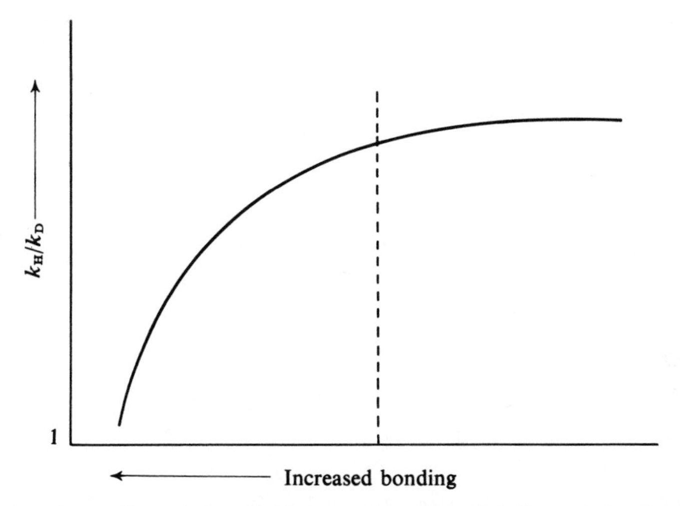

Fig. 5.2 A possible relationship between the value of k_H/k_D and the degree of bonding in the transition state.

5.2.4. The isolation of bridged intermediates. Although product and kinetic studies provide good evidence for the intermediacy of bridged compounds, more convincing evidence comes from the isolation of the bridged intermediates as stable compounds. This has been achieved in a number of favourable cases. The acetolysis of *trans*-2-acetoxycyclohexyl tosylate (108) shows an enhanced rate of reaction and gives the *trans*-diacetate (§5.2.2). Both observations are consistent with the assumption that the acetolysis occurs via the bridged intermediate (109), and this species has been prepared, as the tetrafluoroborate salt (117), by treating *trans*-2-acetoxycyclohexyl bromide (116) in nitromethane with silver tetrafluoroborate. When (117) was treated with anhydrous acetic acid, containing potassium acetate, potassium fluoroborate was precipitated quantitatively, and on keeping the solution at 75 °C for a time the *trans*-diacetate (118) was obtained (Anderson, Freidrich and Winstein, 1963).

The formation of a bridged intermediate has also been demonstrated in the methanolysis of 2-(*p*-hydroxyphenyl)ethyl bromide (119) (Baird and Winstein, 1963). When (119) was treated with excess sodium methoxide in methanol, the initial u.v. absorbance of the reaction solution,

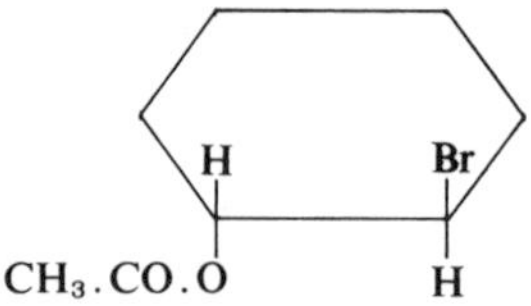

(116) 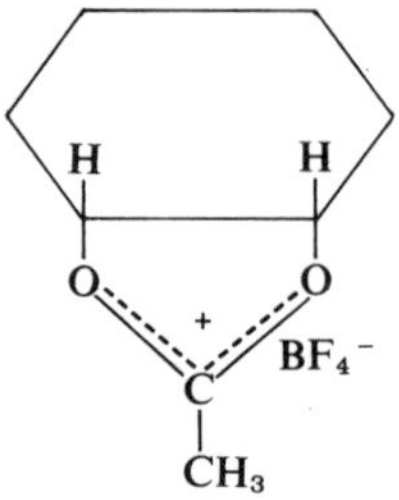 (117) 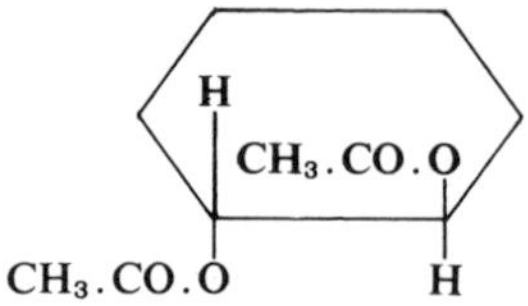 (118)

at 229.5 nm, disappeared and was replaced by an absorbance at 274 nm. After a while this new absorbance began to disappear and was replaced by a new phenolic absorbance at 224 nm. These changes were attributed

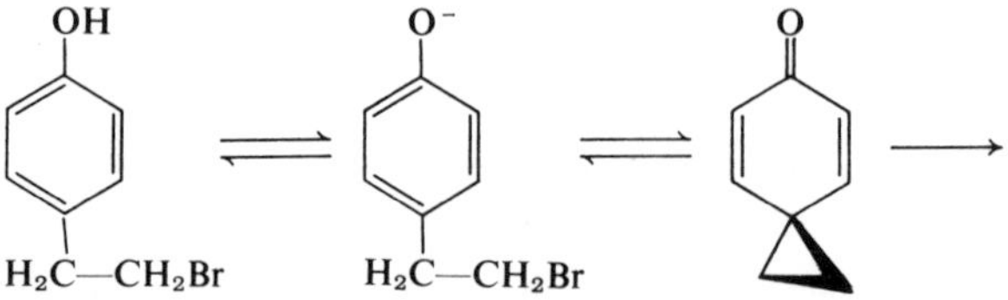

(119) (120) 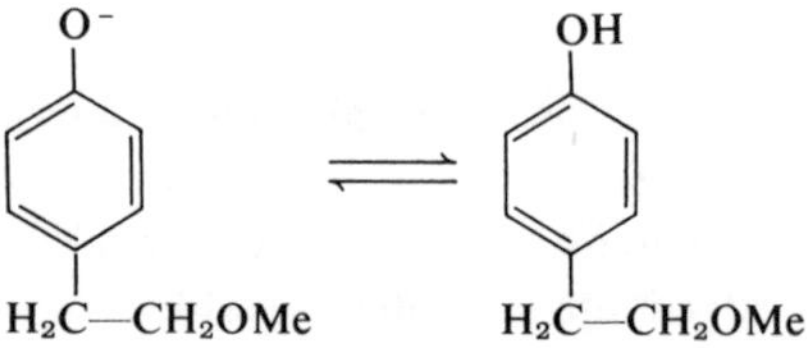 (121)

to the conversions (119)→(120)→(121), in which the intermediate giving the absorbance at 274 nm was assumed to be the spiro-dienone (120). By passing an ethereal solution of (119) over a column of alumina on which was absorbed a strong solution of potassium hydroxide, a sample of (120) was prepared as a pure crystalline solid whose u.v. spectrum in methanol was identical with that of the intermediate formed during the methanolysis. It was further shown that for a solution of (120) in methanol containing sodium methoxide the rate constant for the disappearance of the absorbance maximum at 274 nm was the same as that observed in the conversion of (119) into (121). These experiments show quite clearly that (120) is an intermediate in the methanolysis of (119).

5.2.5. Nuclear magnetic resonance studies. Of the physical methods available, n.m.r. spectroscopy has provided the most useful evidence concerning the existence of stable bridged carbonium ions, although the conditions under which such species have been generated are quite different from those in which they are usually postulated to be intermediates. So that whilst the n.m.r. evidence settles the question of whether bridged carbonium ions can be formed, it gives no indication of whether these species are involved as transient intermediates during nucleophilic substitutions.

The solvolysis of 9-anthrylethyl tosylate (122) shows substantial anchimeric acceleration, which has been explained in terms of the for-

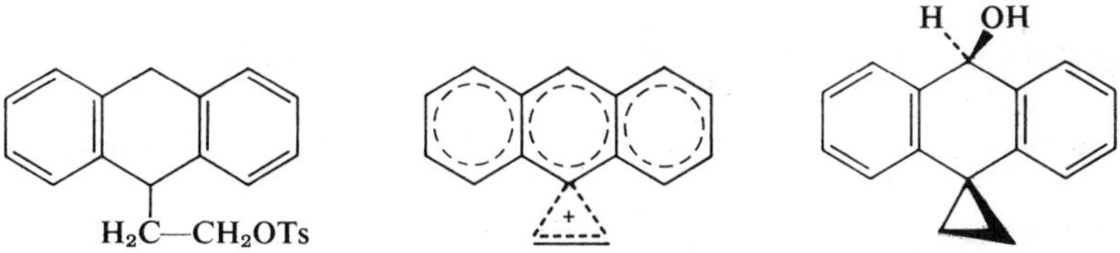

$H_2C—CH_2OTs$

(122) **(123)** **(124)**

H OH

mation of the bridged anthrylethyl cation (123). The bridged carbonium ion has been generated as a stable entity by adding the spirocyclopropyl alcohol (124) to SbF_5–SO_2 at -80 °C (Eberson and Winstein, 1965).

The above n.m.r. evidence indicates that the bridged carbonium ion (123) can be formed, but it does not settle the question of whether such a species is formed by the ionisation of (122). A more useful observation, from this point of view, is that the bridged *p*-anisylethyl cation (126) can be generated by the ionisation of 2-(*p*-anisyl)ethyl chloride (125) in SbF_5–SO_2 at -70 °C (Olah, Comisarow, Namanworth and Ramsey, 1967). Other bridged arylonium ions have also been generated under similar conditions.

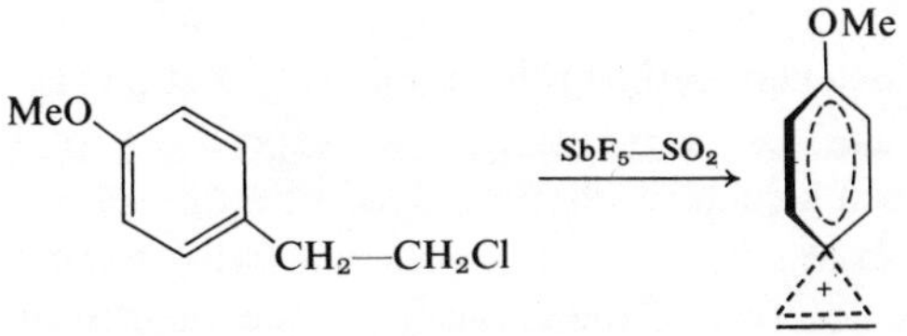

MeO, $CH_2—CH_2Cl$ $\xrightarrow{SbF_5—SO_2}$ OMe

(125) **(126)**

5.3. Reactions of the norbornyl and related systems

In this section we shall consider in more detail some aspects of the solvolysis of 2-norbornyl brosylate and some related compounds. In acetic acid at 25 °C, the rate of acetolysis of *exo*-2-norbornyl brosylate is 350 times greater than that of the *endo*-isomer. The product obtained in both cases is *exo*-2-norbornyl acetate (none of the *endo*-isomer is found), together with a small amount of nortricyclene (127). When optically

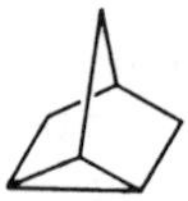

(127)

active starting material is used the product is found to be completely racemic, and the rate of loss of optical activity is greater than the titrimetric rate for the *exo*-compound, but not for the *endo*-isomer. These results can be explained in terms of the reaction scheme [*5.11*].

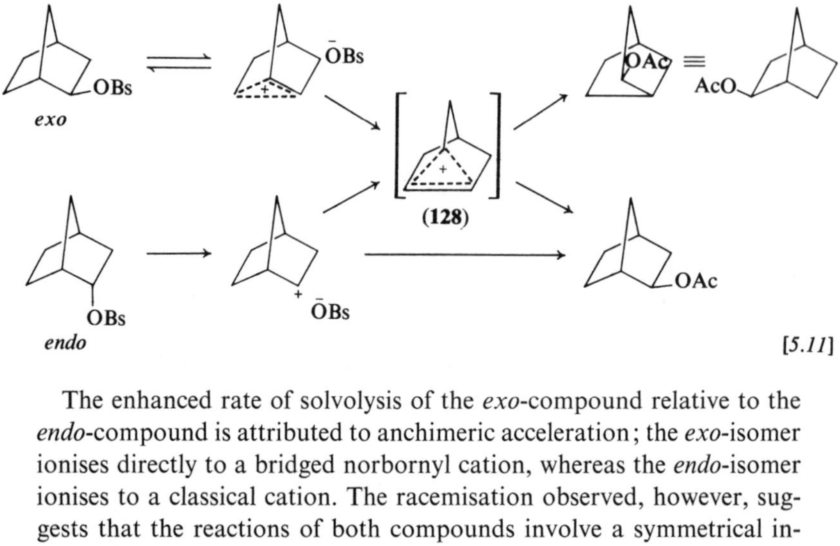

OBs *exo* ⇌ $\bar{O}Bs$ → **(128)** → OAc ≡ AcO

OBs *endo* → $\overset{+}{}$ $\bar{O}Bs$ → OAc

[*5.11*]

The enhanced rate of solvolysis of the *exo*-compound relative to the *endo*-compound is attributed to anchimeric acceleration; the *exo*-isomer ionises directly to a bridged norbornyl cation, whereas the *endo*-isomer ionises to a classical cation. The racemisation observed, however, suggests that the reactions of both compounds involve a symmetrical intermediate, shown here as the bridged norbornyl cation (128), see p. 146. This intermediate also explains the stereospecificity of the reaction, since the bridged structure restricts attack by the solvent to the *exo*-positions. The excess rate of loss of optical activity over rate of solvolysis, observed in the case of the *exo*-isomer, is explained by ion-pair return from the first formed bridged intermediate. The product obtained from the *endo*-isomer shows a slight retention of optical activity which may indicate that some product in this case arises by an attack of the solvent on the first ion-pair intermediate, as indicated in scheme [*5.11*].

The explanation given in the preceding paragraph contains several assumptions that it will be instructive to consider in more detail. It should be pointed out that the fact that only *exo*-product is formed is not, by itself, proof that a bridged intermediate is involved, it merely indicates that the attack by the solvent on the reaction intermediate occurs much more readily at the *exo*- than at the *endo*-position. A direct measure of the ratio of *exo*- to *endo*-attack has been obtained from measurements of the rate of the acid catalysed loss of optical activity of *exo*-2-norbornyl acetate (5.2) and the rate of isomerisation of *exo*-2-norbornyl acetate into the *endo*-compound, i.e. the forward reaction of (5.3) (Goering and Schewene, 1965). The former rate constant provides

$$(+)\ exo\text{-OAc} \xrightarrow{k_\alpha} (\pm)\ \text{products} \qquad (5.2)$$

$$exo\text{-OAc} \underset{k_{\text{i-endo}}}{\overset{k_{\text{i-exo}}}{\rightleftharpoons}} endo\text{-OAc} \qquad (5.3)$$

a measure of the ionisation of the *exo*-acetate (k_1 in scheme [*5.12*]), and the latter rate constant is related to the various constants in [*5.12*] by the relation $k_{\text{i-exo}} = k_1 k_{-2}/(k_{-1} + k_{-2})$. The quantity k_{-1}/k_{-2},

$$exo\text{-OAc} \underset{k_{-1}}{\overset{k_1}{\rightleftharpoons}} \text{Intermediate} \underset{k_{-2}}{\overset{k_2}{\rightleftharpoons}} endo\text{-OAc} \qquad [5.12]$$

which measures the ratio of *exo*- to *endo*-attack on the intermediate, can be obtained from the experimentally determined values of k_α and $k_{\text{i-exo}}$; $k_{-1}/k_{-2} = (k_\alpha/k_{\text{i-exo}}) - 1$. It was found that capture of the intermediate by the solvent, acetic acid, at 25 °C gave 99.99 per cent of the *exo*-acetate, and this represents a difference of 18 kJ mol^{-1} in the apparent activation energies for capture of the intermediate from the *exo*- and *endo*-direction (fig. 5.3). The energy levels in fig. 5.3 were calculated from the temperature dependencies of k_α, $k_{\text{i-exo}}$, and K_{eq}, the last mentioned constant being the equilibrium constant for (5.3).

The results given in fig. 5.3 may be used to explain the stereospecificity shown in the solvolysis of 2-norbornyl brosylate. Moreoever, since a similar energy diagram may be constructed for reactions of the *exo*- and *endo*-brosylates, the factors responsible for the difference in energies of the *exo*- and *endo*-transition states in fig. 5.3 will also be responsible for the difference in rates of solvolysis. The assumption most frequently made is that bridging lowers the energy of the *exo*-transition state, which

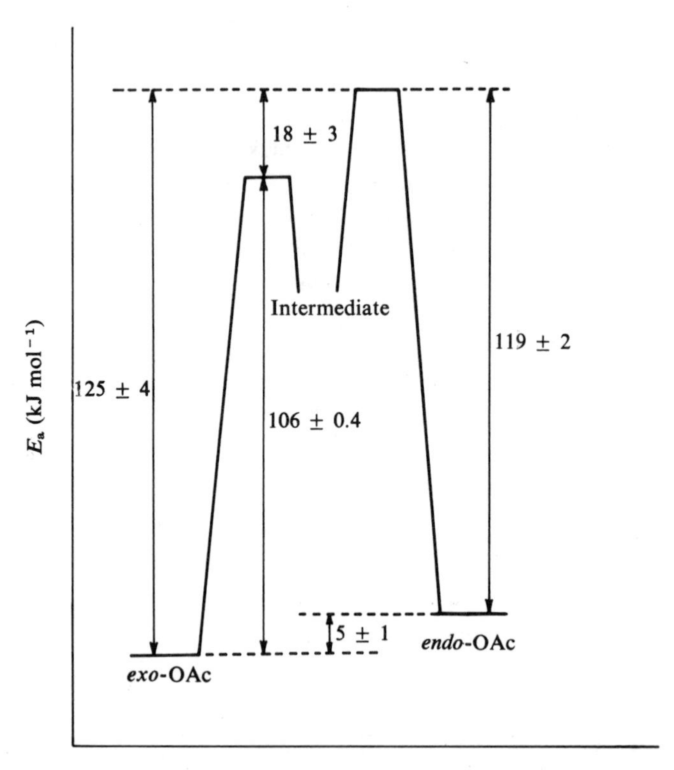

Fig. 5.3 Apparent activation energies for the acid catalysed reactions of *exo*- and *endo*-2-norbornyl acetates in acetic acid at 25 °C. Data from Goering and Schewene (1965).

is therefore more stable than that expected for a classical open carbonium ion. However, the results are also consistent with the assumption that the energy of the *exo*-transition state is normal and that that of the *endo*-transition state is higher than expected because of unfavourable steric interactions (Brown, 1966). In particular the *endo*-hydrogen atom at the C6 position could hinder the departure of the leaving group from the *endo*-position at C2. It therefore becomes important to demonstrate that the high rate of the *exo*-compound is due to anchimeric acceleration resulting from bridging in the transition state.

Several independent pieces of evidence all suggest that the *exo*-compound does, in fact, show an enhanced rate of reaction (Sargent, 1966); just two of these need be mentioned here. The estimation of anchimeric acceleration, using calculated rates for the open ion, indicates (table 5.1) that the *exo*-compound shows an enhanced rate of reaction, whereas the *endo*-compound reacts at about the expected rate. The other piece of evidence comes from measurements of secondary deuterium isotope effects. Although α-, β-, and γ-D effects have been measured, the results are not straightforward to interpret, because scrambling of the deuterium occurs during the solvolysis of the *exo*-compound. However, the observed isotope effects seem to support the idea of bridging in the case of the *exo*-compound (Sunko and Borčić, 1970). The γ-D effect provides particularly striking evidence for a direct interaction between the C6 and C2 positions in the transition state for solvolysis of the *exo*-compound. In the acetolysis of *endo*-2-norbornyl

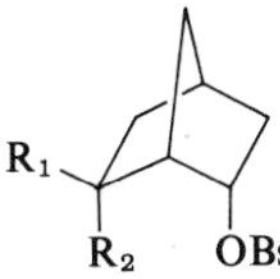

R_1 R_2 OBs **(129)** R_1 OBs R_2 **(130)**

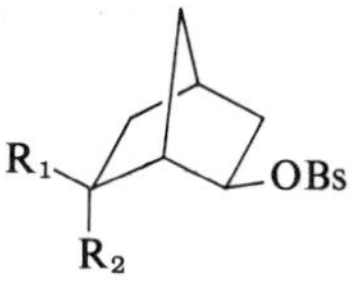

a, R_1 = H, R_2 = D
b, R_1 = D, R_2 = H

brosylate at 65 °C the γ-effect was found to be $k_H/k_D = 1.00$ for (129*a*), and 1.02 for (129*b*). For acetolysis of the *exo*-isomer at 44.4 °C the γ-effect was $k_H/k_D = 1.10$ for (130*a*), and 1.15 for (130*b*). These large isotope effects observed with (130) are strong evidence for bridging in the transition state of the solvolysis reaction, cf. §3.4.4.

5.3.1. The nature of the solvolysis intermediate. The symmetrical intermediate, required in the solvolysis of *endo*- and *exo*-2-norbornyl brosylate, was represented in [*5.11*] by a bridged 2-norbornyl cation. Whilst this structure is sufficient to explain the stereochemical results, it is apparently unable to accommodate the results of experiments using isotopic labelling. The acetolysis of *exo*-2-norbornyl brosylate labelled with ^{14}C at positions C2 and C3 gave product in which the label was distributed as shown in table 5.3 (Robert, Lee and Saunders, 1954).

TABLE 5.3 *Isotopic distributions in the product of acetolysis of exo-2-norbornyl[2,3-$^{14}C_2$] brosylate at 45 °C*

	^{14}C Activity (%)			
	(C2 + C3)	(Cl + C4)	C7	(C5 + C6)
Observed[a]	40	23	22	15
Calculated for:				
(i) Bridged ion (131)	50	25	25	0
(ii) Nortricyclonium ion (132)	33	17	17	33
(iii) 55% (131) + 45% (132)[a]	42	21	21	15
(iv) 6 → 2 (6 → 1) and 3 → 2 Hydride shifts[b]	42.2	22.1	21.1	14.7

[a] Roberts, Lee and Saunders (1954).
[b] Lee and Lam (1966); see text, and footnote *b* table 5.4.

The appearance of the label at positions C5 and C6 cannot be explained in terms of a single intermediate such as (131), which would be expected to give product in which the label was distributed equally between positions Cl and C2, and equally between C3 and C7 [*5.13*].

7 4 5 3 OBs 6 1 2 (131) OAc OAc ≡ AcO [*5.13*]

The observed ^{14}C scrambling was shown to be consistent with the assumption that 55 per cent of the product came from the bridged ion (131) and that 45 per cent came from the more symmetrical intermediate (132) (table 5.3). The latter species, the nortricyclonium ion, was proposed in order to explain the 6 → 2(6 → 1)† hydride shifts that have to occur if the ^{14}C label is to be found at positions C5 and C6. These hydride shifts may, however, occur rapidly [*5.14*], in competition with

† In the bridged norbonyl cation the Cl and C2 positions are equivalent.

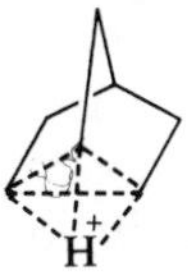

(132)

capture of the bridged norbornyl cation by the solvent, so that it is not necessary to assume that (132) is an intermediate for the solvolysis.

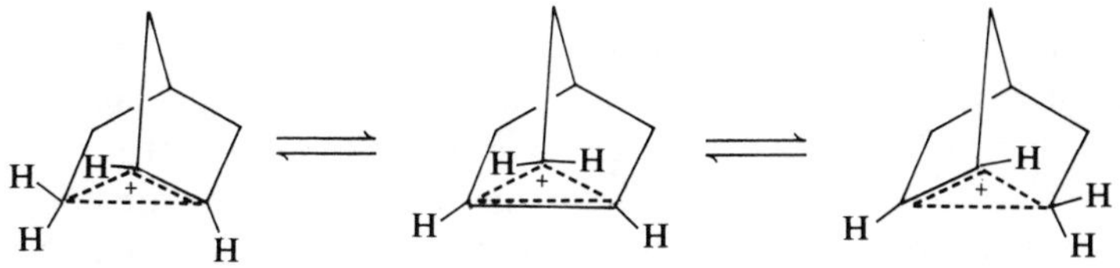

[5.14]

Additional information about the hydride shifts occurring in the solvolysis intermediate comes from the results of scrambling experiments in which a tritium label was used (Lee and Lam, 1966). In the acetolysis of *exo*-2-norbornyl[2-$^3H_1$] brosylate (133), the label was found to be distributed in the product acetate as shown in table 5.4. The observation of activity at C5 and C6 indicates that a 6 → 2 (6 → 1) hydride

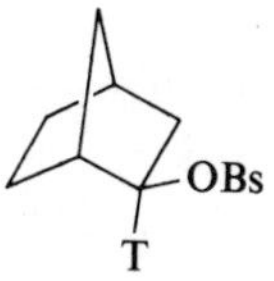

(133)

TABLE 5.4 *Isotopic distributions in the product of acetolysis of exo-2-norbornyl*[2-$^3H_1$] *brosylate at 45 °C*[a]

	^3H Activity (%)			
	C2	C3	(C1 + C4 + C7)	(C5 + C6)
Observed	38.3	1.3	40.1	20.2
Calculated[b]	37.6	1.3	39.6	21.6

[a] Data from Lee and Lam (1966).
[b] Calculated assuming capture of intermediate: (i) before hydride shift (55 per cent), (ii) after 6 → 2 shift (38 per cent), (iii) after 6 → 2 and 3 → 2 shifts (7 per cent).

shift occurs, but the small amount of activity found at C3 indicates that a 3 → 2 hydride shift also takes place.

The observed tritium distribution was accounted for by assuming that hydride shifts occur in the bridged norbornyl cation competitively with capture by the solvent. The major part of the product (55 per cent) apparently arises by capture of the intermediate cation before a hydride shift occurs, a significant part (38 per cent) by capture after a 6 → 2 (6 → 1) hydride shift, and a small amount (7 per cent) by capture after both a 6 → 2 (6 → 1) and a 3 → 2 hydride shift (table 5.4). These proportions also account very nicely for the ^{14}C results. The solvolysis intermediate is thus most reasonably assumed to be a bridged norbornyl cation, in which hydride shifts may occur before capture by the solvent.

The above ideas receive support from n.m.r. studies of the bridged norbornyl cation, which has been generated as a stable species by a variety of methods, for example by dissolving *exo*-2-norbornyl chloride in SbF_5–SO_2 at low temperatures (Olah, White, DeMember, Commeyras and Lui, 1970). The ^1H n.m.r. spectrum is temperature dependent and shows the following changes:

(*a*) in the temperature range +20 °C to −50 °C a single peak is observed;

(*b*) at −80 °C, three groups of peaks having the relative areas 1:4:6 are observed;

(*c*) at −154 °C, the low-field peak of four protons separates into two peaks of two protons each and the high-field peak of six protons broadens and develops a shoulder. The peak of one proton remains unaffected.

These changes can be explained in terms of the gradual freezing out of hydride shifts. When a single peak is observed all protons are equivalent because both 6 → 2 (6 → 1) and 3 → 2 hydride shifts are fast with respect to the n.m.r. time scale, i.e. $H_A \equiv H_B \equiv H_C$ (134). The three peaks

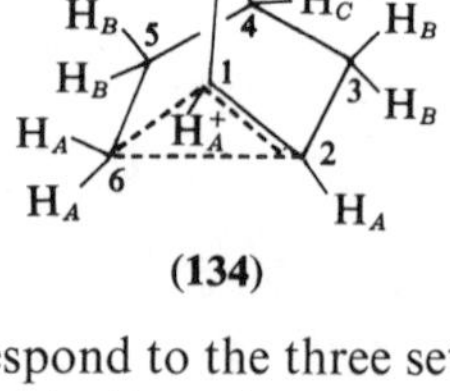

(134)

observed at −80 °C correspond to the three sets of protons H_A, H_B and H_C which become distinguishable owing to the freezing out of the 3 → 2 hydride shift; the remaining equivalences are due to a 6 → 2 (6 → 1) hydride shift. At the lowest temperature studied, the spectrum observed is that of the bridged ion in which all hydride shifts have been frozen out. The only remaining equivalences are the protons attached to C1 and C2, and those attached to C3 and C7. The n.m.r. results thus provide evidence that both 6 → 2 (6 → 1) and 3 → 2 hydride shifts occur in the bridged norbornyl cation, and that the former are more rapid than the latter.

5.3.2. The 7-norbornenyl system. The rate of acetolysis of *anti*-norborn-2-en-7-yl tosylate (136) is about 10^{11} times greater than that of 7-norbornyl tosylate (135). This large difference in rates can be explained by anchimeric acceleration, produced by the participation of the π-bond (137) in the ionisation of (136). A bridged intermediate is also indicated

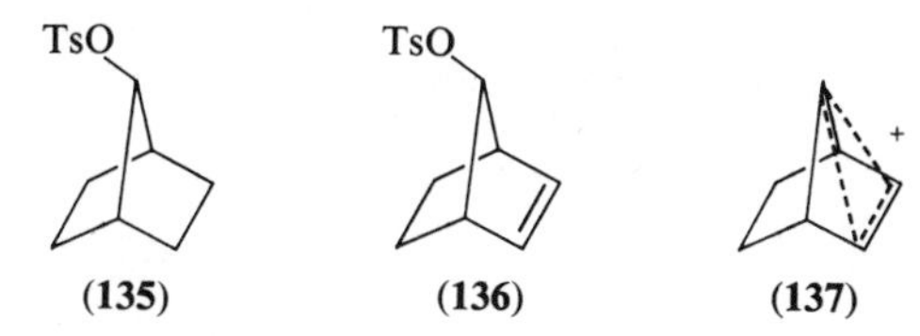

(135) **(136)** **(137)**

by the stereospecificity of the solvolysis, only the *anti*-acetate being produced.

The large difference in the rates of solvolysis of (135) and (136) can be reduced considerably by replacing the hydrogen atom at position C7 by an aryl group. Thus in the solvolysis of the *p*-nitrobenzoates (138) and (139) in 70 per cent aqueous dioxane, the relative rate is reduced to a little over 40 when the aryl group is phenyl (Gassman and Fentiman,

p-NBO Y *p*-NBO Y

(138) **(139)**

1970). Evidently participation by the π-bond becomes less important when the carbonium-ion centre is stabilised by an aryl substituent. By introducing suitable substituents into the ring, π-participation may be apparently eliminated altogether.

The rate constants for solvolysis in 70 per cent aqueous dioxane of the 7-aryl-7-norbornyl compounds (138) are well correlated with the σ^+-values of the ring substituents (fig. 5.4). With the *syn*-7-aryl-*anti*-norborn-2-en-7-yl compounds (139) the $\rho\sigma^+$ correlation shows a change of

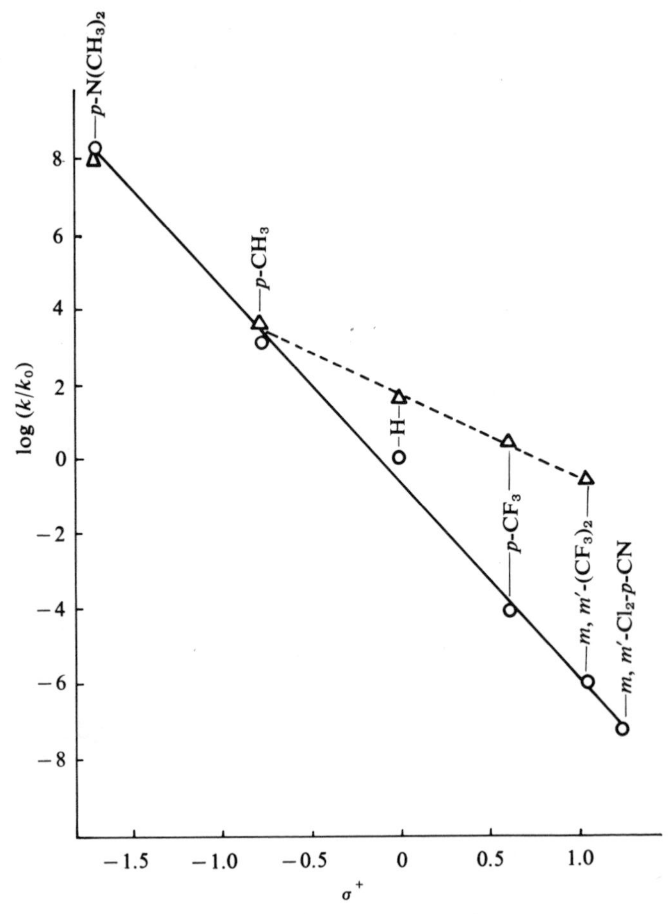

Fig. 5.4 $\rho\sigma^+$ plots for the solvolysis of 7-aryl-7-norbornyl (○) and *syn*-7-aryl-*anti*-norborn-2-en-7-yl *p*-nitrobenzoates (△) in 70 per cent aqueous dioxane at 25 °C. Data from Gassman and Fentiman (1970).

slope at the point for the *p*-anisyl compound (fig. 5.4). These results indicate that the *p*-dimethylaminophenyl and *p*-anisyl groups at the C7 position so stabilise the carbonium-ion intermediate that no participation by the π-bond occurs. The other groups provide less stabilisation, and π-participation becomes important. The change in mechanism implied by the change in slope of the $\rho\sigma^+$ correlation is also consistent with product studies. The compounds that are thought to react by π-participation give stereospecifically the *anti*-product, whereas the other compounds give some of the *syn*-product.

5.4. Reactions of the 3-aryl-2-butyl and related systems

Some of the most convincing evidence for the importance of the phenonium ion intermediate comes from stereochemical studies in the 3-aryl-2-butyl series (Cram, 1964). The parent system, 3-phenyl-2-butyl, possesses two asymmetric carbon atoms and therefore has two sets of diastereoisomers, *erythro* and *threo*, each being an enantiomeric pair [*5.15*]. The *erythro* configuration is that in which the like atoms or

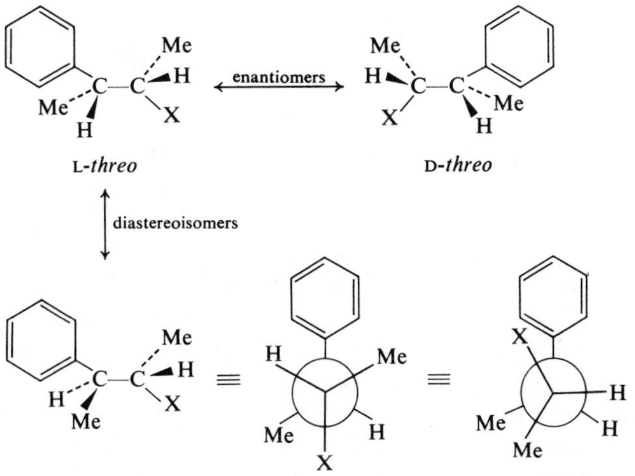

L-*threo* ← enantiomers → D-*threo*

↕ diastereoisomers [*5.15*]

L-*erythro* ≡ ≡ **(140)**

groups on the two carbon atoms may be placed in an eclipsed conformation (140).

The acetolysis of L-*threo*-3-phenyl-2-butyl tosylate at 75 °C gives mainly racemic *threo*-3-phenyl-2-butyl acetate, together with olefinic products and a few percent of the *erythro* diastereoisomer. In contrast, the acetolysis under the same conditions of the D-*erythro*-compound gives mainly D-*erythro*-acetate, together with some olefinic products and a little of the *threo* diastereoisomer. These results can be explained very simply by assuming that a phenonium ion is involved as an intermediate [*5.16*].

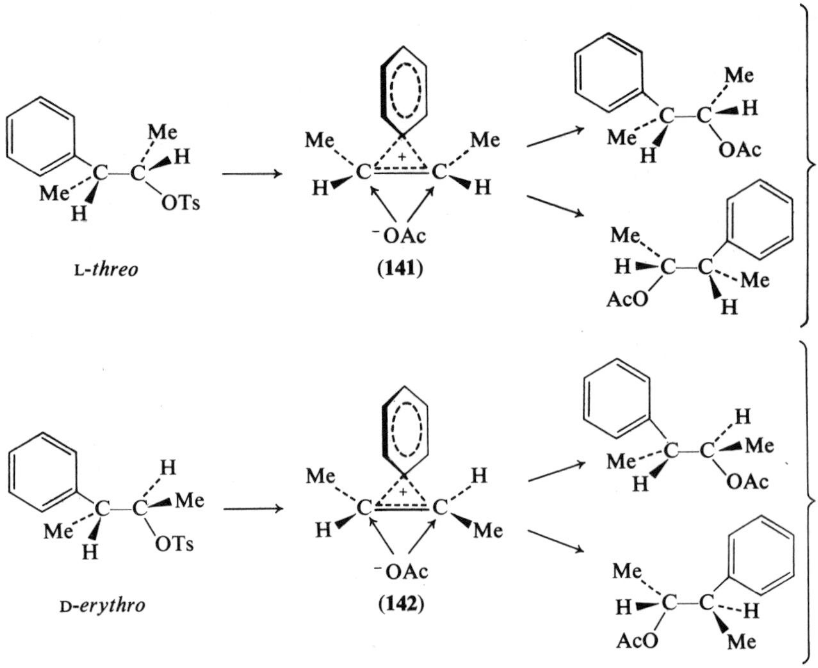

L-*threo*

(141)

racemic *threo*

[*5.16*]

D-*erythro*

(142)

D-*erythro*

The bridged structure of the intermediates restricts the solvent attack to one side of the bridged ions; the plane of symmetry in the case of (141) means that racemic acetate is formed, but optical activity is retained in product arising from (142). The bridged structure also restricts rotation about the carbon–carbon bond, which explains why so little interconversion occurs between the *threo* and *erythro* series. According to [*5.16*], the *threo*-acetate with the inverted configuration is formed by migration of the phenyl group. This has been shown to be the case by using material labelled with ^{14}C at the 1-carbon atom; after solvolysis the racemic acetate was resolved, and the inverted product was shown to be rearranged by phenyl migration [*5.17*] (Smith and Showalter, 1964).

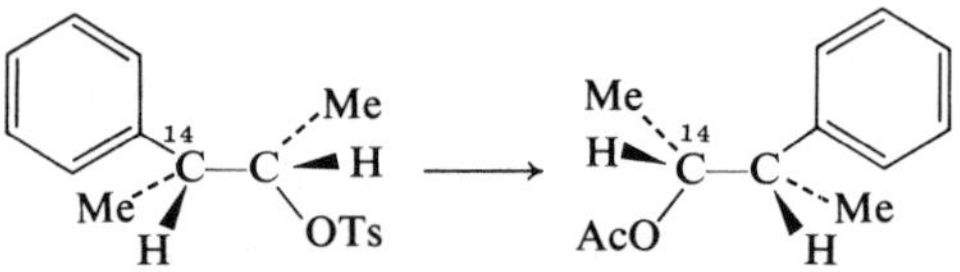

[*5.17*]

Although the hypothesis of a phenonium ion affords a simple rationalisation of the stereochemical results, it has been pointed out that they might be explained alternatively in terms of rapidly equilibrating open ions [*5.18*] (Brown, Morgan and Chloupek, 1965). If such species

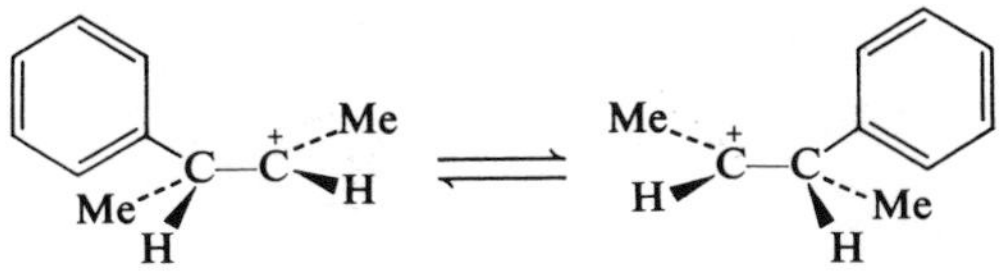

[*5.18*]

are involved, it is necessary, in order to explain the experimental results, to assume that the two open ions equilibrate faster than either reacts with the solvent, and that reaction with the solvent occurs faster than any conformational changes. It has been argued that the rapid movement of the phenyl group would tend both to prevent the solvent from attacking the same face of the carbonium ion as that containing the migrating group, and to restrict rotation about the carbon–carbon bond.

In principle a distinction between the bridged ion and equilibrating open ions may be made from kinetic studies, since anchimeric acceleration is to be expected in the former case, but not in the latter. An accelerated rate arising from ionisation to an open (and more stable) rearranged ion is precluded in the present example, because the rearranged ion is the enantiomer of the unrearranged ion. The practical difficulty of making such a distinction arises because of the uncertainty involved in estimating small rate enhancements (§5.2.2). It should be remembered that the formation of product via a bridged intermediate need not necessarily be accompanied by any appreciable anchimeric acceleration. Of the methods that are used to estimate rate enhancements, that based on a Hammett correlation appears to be one of the most reliable (fig. 5.5). It is assumed that the data for those substrates which contain a deactivating substituent in the phenyl ring establish the correlation line for acetolysis in the absence of a pathway involving aryl participation, i.e. for these substrates $k_t = k_s$ (§5.2.2). Deviations from this line are taken to indicate that the substrate reacts, in part, by a pathway which involves aryl participation, cf. [*5.10*], and the ratio k_t/k_s provides a measure of the anchimeric acceleration. For the parent 3-phenyl-2-butyl system the acceleration so determined is a factor of 3. This type of correlation will be considered further in the following sections.

Part of the evidence supporting the assumption that with electron withdrawing substituents in the phenyl ring reaction occurs largely by the

k_s pathway comes from the loss of stereochemical control during solvolysis of such substrates. Thus in the acetolysis of L-*threo*-3-(*p*-nitrophenyl)-2-butyl tosylate, the secondary acetate obtained was found to be 7 per cent *threo* and 93 per cent *erythro*, in marked contrast to the behaviour

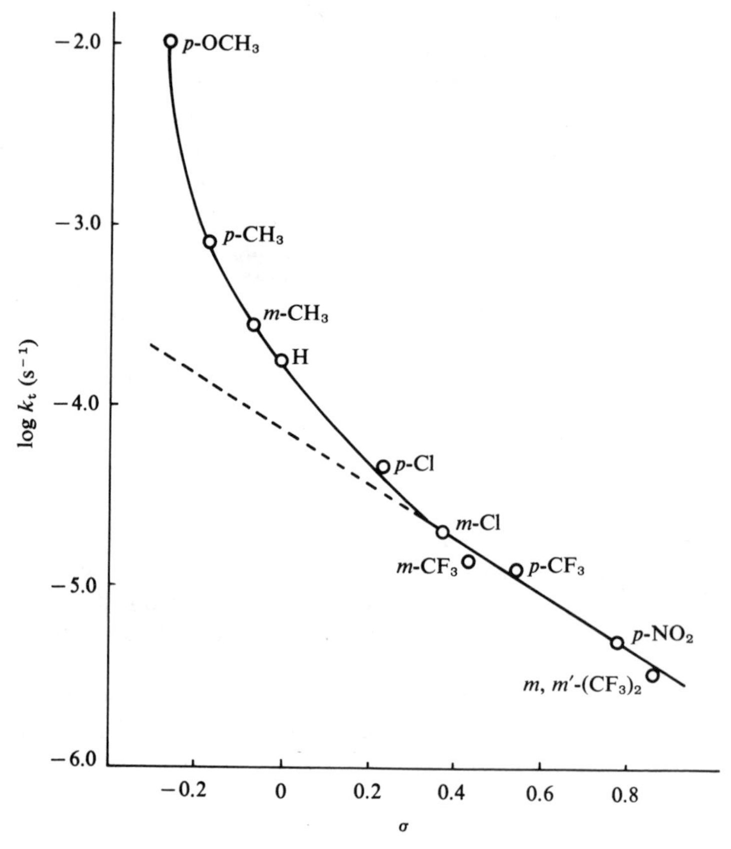

Fig. 5.5 Rates of acetolysis of *threo*-3-aryl-2-butyl brosylates at 75.0 °C against σ-constants. Data from Brown and Kim (1971).

shown by the parent compound [*5.16*]. Presumably, the substituent so reduces the nucleophilicity of the phenyl ring that the internal nucleophile can no longer compete effectively with the external nucleophile, i.e. the k_Δ pathway becomes much less important than the k_s pathway. A similar trend should be produced by changing the solvent, the k_s pathway should be promoted by the more nucleophilic solvent. Finally the k_Δ pathway might become unimportant if the substrate can form a stable open ion, as appears to be the case with some tertiary systems and the secondary, 1,2,2-triphenylethyl system (143) (Cram, 1964).

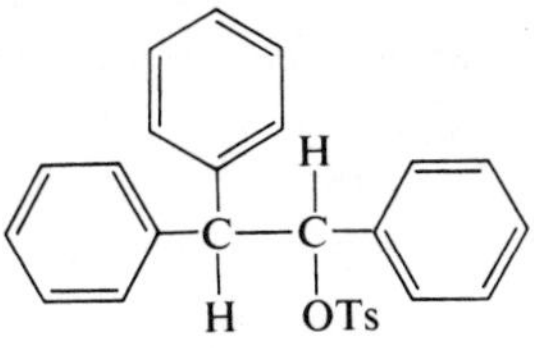

H
C—C
H OTs
(143)

5.4.1. Competing k_Δ and k_s pathways. The solvolysis reactions of 2-phenylethyl arenesulphonate esters can be described in terms of a competition between two reaction pathways [*5.19*]. An essential feature

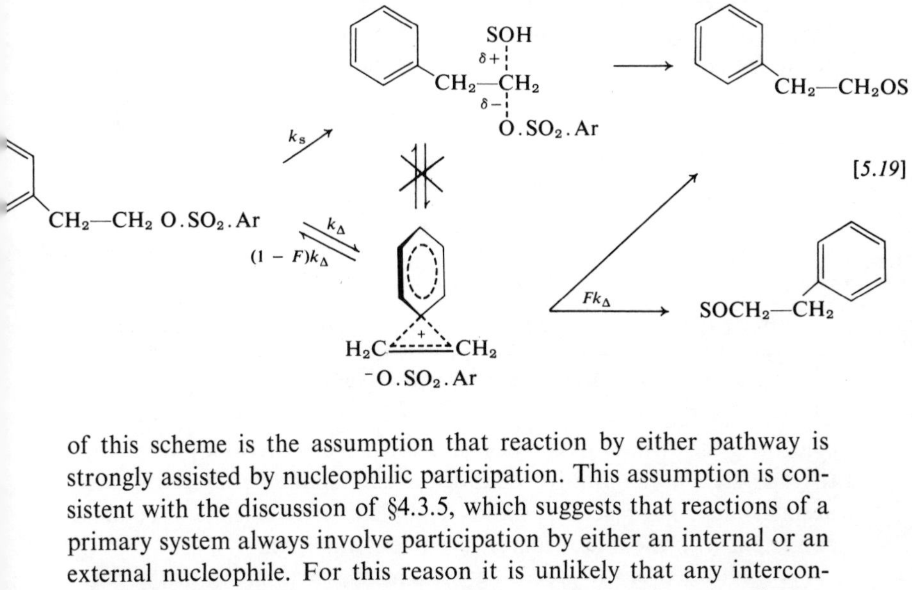

SOH
δ+
CH_2—CH_2
δ−
$O.SO_2.Ar$
CH_2—CH_2OS
k_s
[*5.19*]
CH_2—CH_2 $O.SO_2.Ar$
k_Δ
$(1 - F)k_\Delta$
Fk_Δ
$SOCH_2$—CH_2
H_2C + CH_2
$^-O.SO_2.Ar$

of this scheme is the assumption that reaction by either pathway is strongly assisted by nucleophilic participation. This assumption is consistent with the discussion of §4.3.5, which suggests that reactions of a primary system always involve participation by either an internal or an external nucleophile. For this reason it is unlikely that any interconversion between the two reaction pathways will take place, i.e. discrete processes are involved.

The k_s pathway will most probably have the characteristics of an S_N2 reaction and the structure written as an intermediate is more likely to be a transition state. The k_Δ pathway will involve the formation of a bridged carbonium-ion intermediate, written here as an ion pair. Internal return,

$(1-F)k_\Delta$, from this ion pair is included to explain the ^{14}C-equilibration that occurs in the starting material during the solvolysis of substrate initially labelled at the 1-carbon atom; this labelling also enables the extent of rearrangement in the product of solvolysis to be determined. The fraction of ion pair that proceeds to product is F, and the observed titrimetric rate constant is given by $k_t = k_s + Fk_\Delta$. The value of k_t may be determined directly; from this, the rate of ^{14}C-scrambling in the starting material, and the amount of rearrangement in the product, it is possible to evaluate the various constants in scheme [*5.19*] (Jones and Coke, 1969).

An analysis of this type enables the contributions of the two pathways to be estimated for any reaction conditions. Since the k_s and k_Δ pathways are in competition it is to be expected that the partitioning of a given reaction between the two will depend upon the nature of the solvent. This is borne out by the results included in table 5.5, which show that the solvolysis of 2-phenylethyl tosylate occurs predominantly by the k_s pathway in ethanol, but predominantly by the k_Δ pathway in trifluoroacetic acid. The latter result is supported by stereochemical

TABLE 5.5 *Values of the rate constants k_Δ and k_s for the solvolysis of 2-phenylethyl tosylate at 75 °C*[a]

Solvent	$10^6\ k_\Delta\ (s^{-1})$	$10^6\ k_s\ (s^{-1})$	k_Δ/k_s
EtOH	0.042	7.04	0.006
AcOH	0.215	0.186	1.2
HCOOH	35.4	3.94	9
CF_3COOH	1160	0.05	23000

[a] Data from Diaz, Lazdins and Winstein (1968*b*).

evidence; the trifluoroacetolysis of *threo*-1,2-dideuterio-2-phenylethyl tosylate gives a quantitative yield of the corresponding *threo*-trifluoroacetate, i.e. the reaction shows complete stereochemical control, consistent with the intermediacy of a phenonium ion.

As mentioned in the preceding section, the partitioning of the reaction between the two pathways will also be affected by substituents in the phenyl ring, electron withdrawing substituents tending to reduce the importance of the k_Δ pathway. The correlation observed between the titrimetric rate constants for the acetolysis of a series of 2-arylethyl tosylates and the Hammett σ-values of the ring substituents is shown in fig. 5.6 (Harris, Schadt and Schleyer, 1969). The extrapolation of the linear part of the correlation provides the values of k_s for those substrates that react with anchimeric assistance, i.e. in part by the k_Δ pathway, and for these substrates $(k_t - k_s) = Fk_\Delta$. Thus the contribution made by each pathway may be determined by a purely kinetic method, and the estimate of the amount of aryl participation $(100Fk_\Delta/k_t)$ obtained by this analysis is in very good agreement with that obtained independently

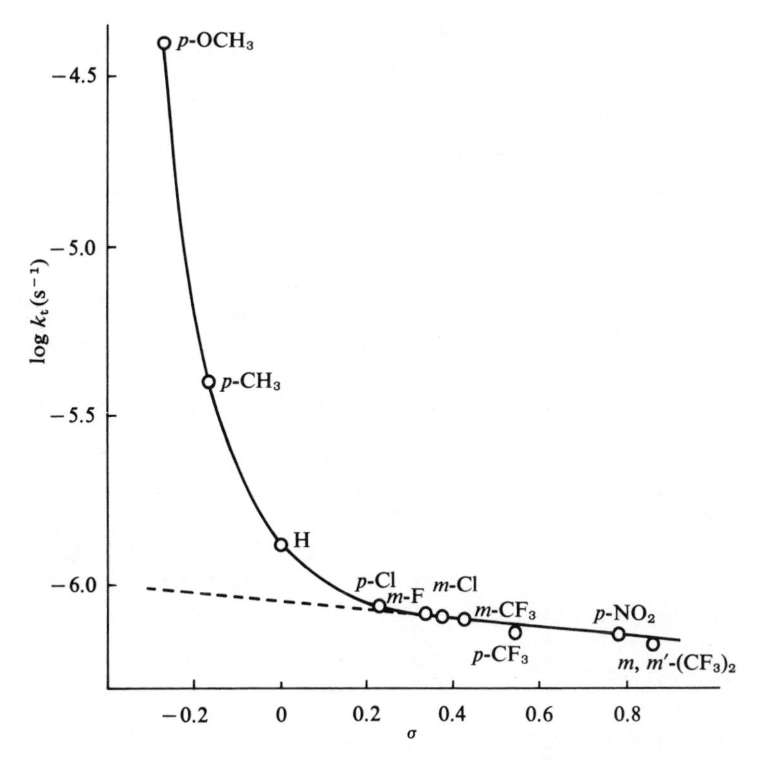

Fig. 5.6 Rates of acetolysis of 2-arylethyl tosylates at 90 °C against σ-constants. Data from Harris, Schadt and Schyler (1969).

from product studies (table 5.6). The agreement between the two estimates implies that a quantitative relationship exists between the product and kinetic data, which is seen as strong evidence in support of the idea that the k_Δ and k_s pathways are discrete processes. If cross-over between the pathways in [*5.19*] occurred, there would not necessarily be a correlation between the amount of rearranged product determined by ^{14}C-scrambling and that calculated by the kinetic analysis.

TABLE 5.6 *Estimates of aryl participation (100 Fk_Δ/k_t) in the acetolysis of 2-arylethyl tosylates*

		p-Cl	H	p-Me	p-MeO
$100Fk_\Delta/k_t$	I[a]	18	42	82	98
	II[b]	8.7	38	82	95

[a] Hammett correlation, data from Harris, Schadt and Schleyer (1969).
[b] ^{14}C-scrambling, data from Jones and Coke (1969).

5.4.2. Correlations based on the Hammett equation. A product-rate correlation of the type described in the preceding section has been observed for the acetolysis of a series of *threo*-3-aryl-2-butyl brosylates (Brown and Kim, 1971). The amount of product arising from the k_Δ pathway was calculated from the kinetic data, using the Hammett plot (fig. 5.5) to evaluate the terms k_s and Fk_Δ, and this was found to correspond closely with the amount of *threo*-acetate determined by product studies (table 5.7). The above correlation is based on the assumption that only

TABLE 5.7 *Products of acetolysis of substituted* threo-*3-phenyl-2-butyl brosylates*[a]

Substituent	Products (%)			*threo*-acetate calculated[b]
	olefins	*erythro*-acetate	*threo*-acetate	
p-MeO[c]	~0.3	0	99.7	99
p-Me	12	0	88	87
m-Me	31	1	68	73
H	38	3	59	66
p-Cl	53	6	39	37

[a] Data for 75 °C from Brown and Kim (1971).
[b] Calculated by 100 Fk_Δ/k_t.
[c] Temperature, 50 °C.

retained substitution product (i.e. *threo*-acetate) is formed by the k_Δ pathway, and that all other products are formed by the k_s pathway. The close agreement between the observed and calculated amounts of *threo*-acetate is seen as support for the view that these represent discrete reaction pathways for the acetolysis, cf. [*5.19*], without any significant interconversion occurring between the two.

Substitution by the k_s pathway with inversion of configuration at the 1-carbon atom would lead to *erythro*-acetate, but as will be seen from table 5.7 the amount of that isomer actually formed is very small, and the products other than the *threo*-acetate are mainly olefinic. The significance of the observed product-rate correlation is therefore uncertain until the origin of the olefinic products has been established. A similar criticism does not apply to the correlation observed with the primary systems, §5.4.1, since in those cases the only products of acetolysis were the 2-arylethyl acetates.

Another aspect of the analysis of the kinetic results, which although justifiable in the case of primary systems is more doubtful in the case of secondary systems, is the implicit assumption when using the equation $k_t = k_s + Fk_\Delta$ that ion-pair return is important only in the k_Δ pathway. The evidence presented in §4.3.4 suggests that a nucleophilic displacement at a secondary carbon atom may involve ionisation to a carbonium-ion intermediate, which may react with solvent, dissociate, or return to starting material. If such an intermediate is involved in the k_s pathway, then the possibility of ion-pair return cannot be excluded. The titrimetric rate constant under these circumstances will be given by $k_t = F'k_s + Fk_\Delta$, in which F' represents the partitioning of the solvolysis intermediate in the k_s pathway. This makes the analysis of the kinetic results more difficult, because an enhanced value of k_t (indicated by curvature in the Hammett plot) will now be consistent with either an increase in the contribution from the aryl assisted pathway, Fk_Δ, or a reduction in the amount of ion-pair return associated with the k_s pathway; less ion-pair return means an increase in the partition factor F'. A product-rate correlation might still be observed (§5.5), although the apparent significance of such a relationship would depend upon the mechanisms assumed for the k_s and k_Δ pathways; these in turn are partly determined by the assumptions that are made about the nature of the nucleophilic participation by the solvent.

5.5. The nature of solvent participation

The assumption that solvolysis may take place by two discrete pathways implies that the reaction co-ordinates for the two pathways are separated by an energy barrier sufficient to prevent any significant cross-over between the two. This requirement is seen to impose a restriction on the nature of the k_s process, *viz.* an open carbonium ion is unlikely to be involved, because there would probably be only a small energy barrier between the open and the corresponding bridged ion so that the two would interconvert rapidly and the discreteness of the pathways would be lost.

This difficulty may be overcome by assuming that nucleophilic participation is an important factor, not only in the k_Δ pathway but also in the k_s pathway. In the former case the nucleophilic participation is due to an interaction between the reacting centre and a neighbouring group, and in the latter case it is due to an interaction involving a solvent molecule; the transition states for the two pathways may be represented as (144) and (145). It is unlikely that interconversion between species

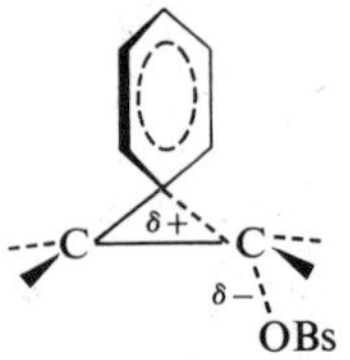

(144) 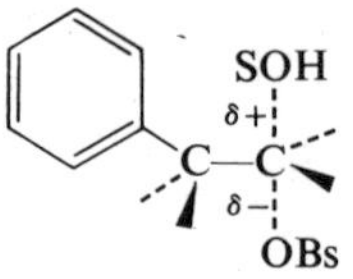 **(145)**

involving strong participation occurs very easily, which satisfies the necessary condition for obtaining discrete reaction pathways and for observing a product-rate correlation.

The conclusion that solvent participation is important in the k_s pathway applies to all those systems for which a product-rate correlation has been observed. The results of the preceding sections therefore seem to suggest that solvent participation must be an important factor in the solvolysis of simple secondary systems as well as of primary systems. Depending upon the assumptions made about the nature of solvent participation this may or may not require some modification to be made to current views about the mechanisms of solvolysis of secondary systems.

Solvent participation may be represented as a nucleophilic interaction between a molecule of solvent and the reacting carbon atom during ionisation (145), as in the structural hypothesis of Doering and Ziess (§1.2.3). Such participation is expected to stabilise the transition state for ionisation and so lead to an enhanced rate of solvolysis, relative to that of a process which involves ionisation without participation. Evidence to support this view comes from the difference in behaviour of 2-adamantyl tosylate (146) and isopropyl tosylate. The relative rate of solvolysis, $k_{\text{i-Pr}}/k_{\text{2-Ad}}$, in the different solvents (table 5.8) shows a large variation and indicates that the two systems behave very differently towards a change in the solvent (Schleyer, Fry, Lam and Lancelot, 1970). Although (146) is a secondary system it appears to react, by all the usual criteria of mechanism, by a limiting S_N1 mechanism, whereas isopropyl tosylate is a typical simple secondary system which exhibits borderline behaviour. The change in relative reactivity is thus seen as a consequence of greater solvent participation by the more nucleophilic solvents in the case of isopropyl tosylate than in the case of 2-adamantyl tosylate.

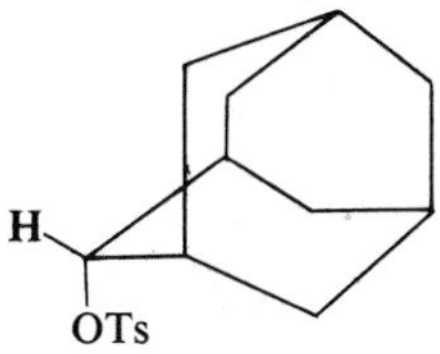

(146)

TABLE 5.8 *The relative rates of solvolysis of some secondary arenesulphonates in different solvents at 25 °C*

Solvent	$k_{\text{i-Pr}}k_{\text{2-Ad}}$[a]	$k_{\text{di-t-Bu}}/k_{\text{2-Ad}}$[b]	$k_{\text{i-Pr}}/k_{\text{Pin}}$[c]
CF_3COOH	0.0056	–	0.00036
97% TFE	1.6[b]	1000	0.026
HCOOH	3.2	1250	0.071
CH_3COOH	12.5	1000	0.344
50% EtOH	20	–	0.714
80% EtOH	125	1000	2.27
90% EtOH	–	–	3.35
100% EtOH	1000	–	–

[a] Data from Schleyer, Fry, Lam and Lancelot (1970).
[b] Data from Liggero, Harper, Schleyer, Krapcho and Horn (1970).
[c] Data from Shiner, Fisher and Dowd (1969).

This difference in behaviour can be explained in terms of the different steric requirements of the two substrates; nucleophilic participation is possible with the isopropyl system (147), but not with the 2-adamantyl system (148), because of the unfavourable non-bonding interactions in

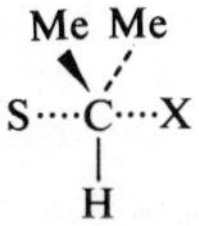

(147) 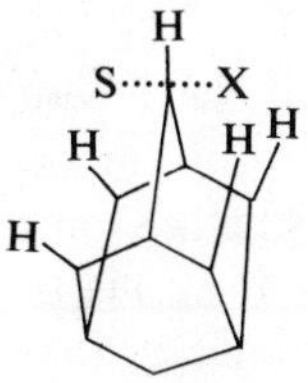 **(148)**

the latter case involving the axial hydrogen atoms and both the incoming and leaving groups. This explanation receives some support from the observation that nucleophilic participation is apparently eliminated when the methyl groups in (147) are replaced by the more bulky t-butyl

$$(CH_3)_3C\text{—}\overset{\displaystyle H}{\underset{\displaystyle X}{C}}\text{—}C(CH_3)_3$$

(149)

groups. The di-t-butylcarbinyl system (149) so obtained shows an almost constant rate of solvolysis relative to that of the 2-adamantyl system (table 5.8) in a wide range of solvents.

The relative rates in the above examples were calculated with solvolysis rate constants, and so may not be reliable indications of participation in ionisation if ion-pair return is important. An enhanced rate of solvolysis could arise if participation after ionisation caused a reduction in the amount of ion-pair return (§5.4.2). The possible magnitude of this effect is indicated by the relative rates $k_{\text{i-Pr}}/k_{\text{Pin}}$ (table 5.8) in which k_{Pin} is the rate constant of solvolysis of 3,3-dimethyl-2-butyl (pinacolyl) brosylate, which has been proposed as a reference compound for estimating unassisted rates of ionisation of secondary brosylates in the absence of ion-pair return (Shiner, Fisher and Dowd, 1969). Using rate constants of ionisation to calculate rate enhancements enables acceleration due to participation in ionisation to be distinguished from acceleration due to participation after ionisation. The results suggest that the variation in the ratio $k_{\text{i-Pr}}/k_{\text{2-Ad}}$ could be due to changes in the amounts of ion-pair return, although in the more nucleophilic solvents (for which $k_{\text{i-Pr}}/k_{\text{Pin}} > 1$) participation in ionisation may also be an important factor. Thus solvent participation can be described in terms of the effect of the solvent on the partitioning of an ion-pair intermediate, without any necessary involvement of the solvent in the ionisation step.

The difference between the two descriptions of solvent participation lies not in the nature of the interactions involved, but in the timing of the various steps of the reaction pathway, in particular whether an interaction with the solvent occurs before or after the ionisation. If participation occurs before ionisation, then the discrete k_Δ and k_s pathways are explained in terms of the discussion presented earlier, and current interpretations of the mechanisms of solvolysis of simple secondary compounds require modification. If, on the other hand, participation occurs after ionisation, the only slight change required to the ion-pair hypothesis is the inclusion of a solvent molecule, e.g. (150), to indicate that specific solvation by a molecule of the solvent is important (cf. Raber, Harris, Hall and Schleyer, 1971).

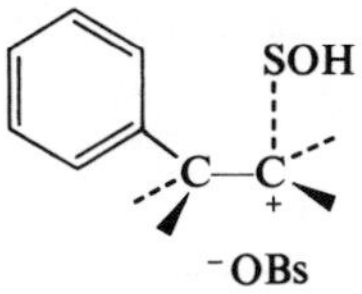

SOH

$^{-}$OBs

(150)

The inclusion of a solvent molecule in an ion-pair intermediate, without necessarily implying its presence in the transition state of the ionisation, can also lead to a mechanism of solvolysis which is consistent with the observation of a product–rate correlation. One such mechanism which satisfies the conditions is the prior formation of a tight ion pair (an open ion formed without participation) followed by the competitive participation by a solvent molecule and by the neighbouring aryl group (Brown and Kim, 1971) [*5.20*]. It can be shown that a

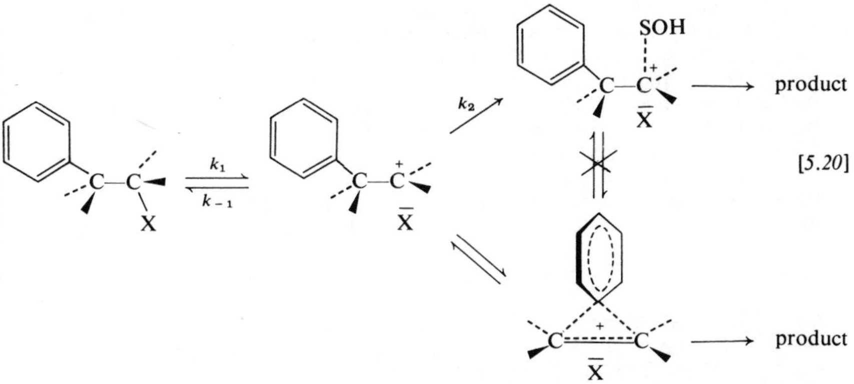

k_1 k_{-1} k_2 SOH product product X [*5.20*]

satisfactory product–rate correlation can be expected for scheme [*5.20*] when $k_{-1}/k_2 > ca\ 20$. The specific solvation of the open ion by a solvent molecule in the k_s pathway is an important feature of the scheme, because it explains why interconversion does not occur between this species and the bridged ion of the k_Δ pathway.

The above discussion is sufficient to show that the observation of a product–rate correlation using a Hammett plot based on solvolysis (titrimetric) rate constants does not provide unequivocal information about the nature of the k_s pathway.

References

Abraham, M. H. (1969). *Chem. Comm.* 1307.
Albery, W. J. and Robinson, B. H. (1969). *Trans. Faraday Soc.* **65**, 980.
Allinger, N. L., Tribble, M. T., Miller, M. A. and Wertz, D. H. (1971). *J. Am. chem. Soc.* **93**, 1637.
Anderson, C. B., Friedrich, E. C. and Winstein, S. (1963). *Tetrahedron Lett.* 2037.
Anh, N. T. (1968). *Chem. Comm.* 1089.
Arnett, E. M., Bentrude, W. G., Burke, J. J. and Duggleby, P. McC. (1965). *J. Am. chem. Soc.* **87**, 1541.
Arnett, E. M., Bentrude, W. G. and Duggleby, P. McC. (1965). *J. Am. chem. Soc.* **87**, 2048.
Atkinson, G. and Kor, S. K. (1965). *J. phys. Chem.* **69**, 128.
Austin, J. M., Ibrahim, O. D. E.-S. and Spiro, M. (1969), *J. chem. Soc.* (B). 669.
Baird, R. and Winstein, S. (1963). *J. Am. chem. Soc.* **85**, 567.
Bartlett, P. D. and Tidwell, T. T. (1968). *J. Am. chem. Soc.* **90**, 4421.
Benfey, O. T., Hughes, E. D. and Ingold, C. K. (1952). *J. chem. Soc.* 2488.
Bethell, D. and Gold, V. (1967). *Carbonium Ions.* London: Academic Press.
Brown, H. C. (1966). *Chem. in Britain.* 199.
Brown, H. C. and Eldred, N. R. (1949). *J. Am. chem. Soc.* **71**, 445.
Brown, H. C. and Ichikawa, K. (1957). *Tetrahedron.* **1**, 221.
Brown, H. C. and Kim, C. J. (1971). *J. Am. chem. Soc.* **93**, 5765.
Brown, H. C., Morgan, K. J. and Chloupek, F. J. (1965). *J. Am. chem. Soc.* **87**, 2137.
Bunton, C. A. (1963). *Nucleophilic Substitution at a Saturated Carbon Atom.* London: Elsevier.
Capon, B. (1964). *Quart. Rev. chem. Soc. Lond.* **18**, 45.
Ceccon, A., Papa, I. and Fava, A. (1966). *J. Am. chem. Soc.* **88**, 4643.
Clark, G. A. and Taft, R. W. (1962). *J. Am. chem. Soc.* **84**, 2295.
Cook, D. and Parker, A. J. (1968). *J. chem. Soc.* (B) 142.
Cowdrey, W. A., Hughes, E. D., Ingold, C. K., Masterman, S. and Scott, A. D. (1937). *J. chem. Soc.* 1252.
Cram, D. J. (1953). *J. Am. chem. Soc.* **75**, 332.
Cram, D. J. (1964). *J. Am. chem. Soc.* **86**, 3767.
Dafforn, G. A. and Streitwieser, A. (1970). *Tetrahedron Lett.* **36**, 3159.
Davis, R. E. (1965). *J. Am. chem. Soc.* **87**, 3010.
Diaz, A. F., Lazdins, I. and Winstein, S. (1968*a*). *J. Am. chem. Soc.* **90**, 1904.
Diaz, A. F., Lazdins, I. and Winstein, S. (1968*b*). *J. Am. chem. Soc.* **90**, 6546.
Doering, W. E. and Zeiss, H. H. (1953). *J. Am. chem. Soc.* **75**, 4733.
Dostrovsky, I. and Hughes, E. D. (1946). *J. chem. Soc.* 166.
Eberson, L. and Winstein, S. (1965). *J. Am. chem. Soc.* **87**, 3506.
Edwards, J. O. (1956). *J. Am. chem. Soc.* **78**, 1819.
Edwards, J. O. and Pearson, R. G. (1962). *J. Am. chem. Soc.* **84**, 16.

Fort, R. C. and Schleyer, P. v. R. (1966). *Advances in Alicyclic Chemistry*. **1**, 283.
Fry, J. L., Harris, J. M., Bingham, R. C. and Schleyer, P. v. R. (1970). *J. Am. chem. Soc.* **92**, 2540.
Gassman, P. G. and Fentiman, A. F. (1970). *J. Am. chem. Soc.* **92**, 2549.
Gilchrist, T. L. and Storr, R. C. (1972). *Organic Reactions and Orbital Symmetry*. Cambridge University Press.
Gleicher, G. J. and Schleyer, P. v. R. (1967). *J. Am. chem. Soc.* **89**, 582.
Goering, H. L. and Hopf, H. (1971). *J. Am. chem. Soc.* **93**, 1224.
Goering, H. L. and Schewene, C. B. (1965). *J. Am. chem. Soc.* **87**, 3516.
Goering, H. L. and Thies, R. W. (1968). *J. Am. chem. Soc.* **90**, 2967.
Gold, V. (1956). *J. chem. Soc.* 4633.
Gregory, B. J., Kohnstam, G., Queen, A. and Reid, D. J. (1971). *Chem. Comm.* 797.
Grunwald, E. and Winstein, S. (1948). *J. Am. chem. Soc.* **70**, 846.
Haberfield, P., Nudelman, A., Bloom, A., Romm, R. and Ginsberg, H. (1971). *J. org. Chem.* **36**, 1792.
Hammett, L. P. (1970). *Physical Organic Chemistry*, 2nd ed. New York: McGraw-Hill.
Harris, J. M., Schadt, F. L. and Schleyer, P. v. R. (1969). *J. Am. chem. Soc.* **91**, 7508.
Hughes, E. D., Juliusberger, F., Masterman, S., Topley, B. and Weiss, J. (1935). *J. chem. Soc.* 1525.
Ibne-Rasa, K. M. (1967). *J. chem. Educ.* **44**, 89.
Ingold, C. K. (1969). *Structure and Mechanism in Organic Chemistry*, 2nd ed. London: Bell.
Jackson, E. and Kohnstam, G. (1965). *Chem. Comm.* 279.
Jefford, C. W., Sweeney, A., Hill, D. T. and Delay, F. (1971). *Helv. Chim. Acta.* **54**, 1691.
Jerkunica, J. M., Borčić, S. and Sunko, D. E. (1967). *J. Am. chem. Soc.* **89**, 1732.
Johnson, C. D. (1973). *The Hammett Equation*. Cambridge University Press.
Jones, M. G. and Coke, J. L. (1969). *J. Am. chem. Soc.* **91**, 4284.
Kice, J. L., Scriven, R. L., Koubek, E. and Barnes, M. (1970). *J. Am. chem. Soc.* **92**, 5608.
Kohnstam, G. (1967). *Adv. Phys. Org. Chem.* **5**, 121.
Kohnstam, G., Queen, A. and Ribar, T. (1962). *Chem. and Ind.* 1287.
Lee, C. C. and Lam, L. K. M. (1966). *J. Am. chem. Soc.* **88**, 2831.
Lee, C. C. and Noszkó, L. (1966). *Canad. J. chem.* **44**, 2491.
Leffler, J. E. and Grunwald, E. (1963). *Rates and Equilibria of Organic Reactions*. New York: Wiley.
Liggero, S. H., Harper, J. J., Schleyer, P. v. R., Krapcho, A. P. and Horn, D. E. (1970). *J. Am. chem. Soc.* **92**, 3789.
Melander, L. (1960). *Isotope Effects on Reaction Rates*. New York: The Ronald Press.
Murr, B. L. and Donnelly, M. F. (1970). *J. Am. chem. Soc.* **92**, 6688.
Murr, B. L. and Santiago, C. (1968). *J. Am. chem. Soc.* **90**, 2964.
Okamoto, K., Kushiro, H., Nitta, I. and Shingu, H. (1967). *Bull. chem. Soc. Japan.* **40**, 1900.
Okamoto, K., Uchida, N., Saito, S. and Shingu, H. (1966). *Bull. chem. Soc. Japan.* **39**, 307.
Olah, G. A., Comisarow, M. B., Namanworth, E. and Ramsey, B. (1967). *J. Am. chem. Soc.* **89**, 5259.
Olah, G. A., White, A. M., DeMember, J. R., Commeyras, A. and Lui, C. Y. (1970). *J. Am. chem. Soc.* **92**, 4627.

Parker, A. J. (1969). *Chem. Rev.* **69**, 1.

Pearson, R. G., Sobel, H. and Songstad, J. (1968). *J. Am. chem. Soc.* **90**, 319.

Perrin, C. L. and Pressing, J. (1971). *J. Am. chem. Soc.* **93**, 5705.

Pocker, Y. and Kevill, D. N. (1965). *J. Am. chem. Soc.* **87**, 4760.

Pocker, Y., Mueller, W. A., Naso, F. and Tocchi, G. (1964). *J. Am. chem. Soc.* **86**, 5012.

Raber, D. J., Bingham, R. C., Harris, J. M., Fry, J. L. and Schleyer, P. v. R. (1970). *J. Am. chem. Soc.* **92**, 5977.

Raber, D. J. and Harris, J. M. (1972). *J. chem. Educ.* **49**, 60.

Raber, D. J., Harris, J. M., Hall, R. E. and Schleyer, P. v. R. (1971). *J. Am. chem. Soc.* **93**, 4821.

Reich, I. L., Diaz, A. and Winstein, S. (1969). *J. Am. chem. Soc.* **91**, 5635.

Reichardt, C. (1965). *Angew. Chem. internat. Edit.* **4**, 29.

Roberts, J. D., Lee, C. C. and Saunders, W. H. (1954). *J. Am. chem. Soc.* **76**, 4501.

Robertson, R. E. (1967). *Prog. Phys. Org. Chem.* **4**, 213.

Robinson, R. A. and Stokes, R. H. (1959). *Electrolyte Solutions*, 2nd ed. London: Butterworths.

Sargent, G. D. (1966). *Quart. Rev. chem. Soc. Lond.* **20**, 301.

Schleyer, P. v. R. (1964). *J. Am. chem. Soc.* **86**, 1854, 1856.

Schleyer, P. v. R., Fry, J. L., Lam, L. K. M. and Lancelot, C. J. (1970). *J. Am. chem. Soc.* **92**, 2542.

Scott, J. M. W. (1970). *Canad. J. chem.* **48**, 3807.

Seltzer, S. and Zavitsas, A. A. (1967). *Canad. J. Chem.* **45**, 2023.

Shiner, V. J. (1970). In *Isotope Effects in Chemical Reactions* (edited C. J. Collins and N. S. Bowman). New York: Van Nostrand Reinhold.

Shiner, V. J., Buddenbaum, W. E., Murr, B. L. and Lamaty, G. (1968). *J. Am. chem. Soc.* **90**, 418.

Shiner, V. J. and Dowd, W. (1969). *J. Am. chem. Soc.* **91**, 6528.

Shiner, V. J. and Dowd, W. (1971). *J. Am. chem. Soc.* **93**, 1029.

Shiner, V. J. and Fisher, R. D. (1971). *J. Am. chem. Soc.* **93**, 2553.

Shiner, V. J., Fisher, R. D. and Dowd, W. (1969). *J. Am. chem. Soc.* **91**, 7748.

Smith, S. G. (1961). *J. Am. chem. Soc.* **83**, 4285.

Smith, W. B. and Showalter, M. (1964). *J. Am. chem. Soc.* **86**, 4136.

Sneen, R. A. and Larsen, J. W. (1969). *J. Am. chem. Soc.* **91**, 362.

Stork, G. and White, G. N. (1956). *J. Am. chem. Soc.* **78**, 4609.

Streitwieser, A. (1956). *J. Am. chem. Soc.* **78**, 4935.

Streitwieser, A. (1962). *Solvolytic Displacement Reactions.* New York: McGraw-Hill.

Streitwieser, A., Hammond, H. A., Jagow, R. H., Williams, R. M., Jesaitis, R. G., Chang, C. J. and Wolf, R. (1970). *J. Am. chem. Soc.* **92**, 5141.

Streitwieser, A., Jagow, R. H., Fahey, R. C. and Suzuki, S. (1958). *J. Am. chem. Soc.* **80**, 2326.

Streitwieser, A. and Walsh, T. D. (1965). *J. Am. chem. Soc.* **87**, 3686.

Streitwieser, A., Walsh, T. D. and Wolfe, J. R. (1965). *J. Am. chem. Soc.* **87**, 3682.

Sunko, D. E. and Borčić, S. (1970). In *Isotope Effects in Chemical Reactions* (edited C. J. Collins and N. S. Bowman). New York: Van Nostrand Reinhold.

Swain, C. G. (1948). *J. Am. chem. Soc.* **70**, 1119.

Swain, C. G. and Scott, C. B. (1953). *J. Am. chem. Soc.* **75**, 141.

Sykes, P. (1972). *The Search for Organic Reaction Pathways.* London: Longman.

Szwarc, M. (1969). *Accounts Chem. Res.* **2**, 87.

Wells, P. R. (1968). *Linear Free Energy Relationships.* London: Academic Press.

Westheimer, F. H. (1956). In *Steric Effects in Organic Chemistry* (edited M. S. Newman). New York: Wiley.

Whiting, M. C. (1966). *Chem. in Britain.* 482.
Winstein, S., Appel, B., Baker, R. and Diaz, A. (1965). *Spec. Publ. chem. Soc.* no. 19, p. 109.
Winstein, S. and Clippinger, E. (1956). *J. Am. chem. Soc.* **78**, 2784.
Winstein, S., Clippinger, E., Fainberg, A. H., Heck, R. and Robinson, G. C. (1956). *J. Am. chem. Soc.* **78**, 328.
Winstein, S. and Fainberg, A. H. (1957). *J. Am. chem. Soc.* **79**, 5937.
Winstein, S., Grunwald, E. and Jones, H. W. (1951). *J. Am. chem. Soc.* **73**, 2700.
Winstein, S. and Heck, R. (1956). *J. Am. chem. Soc.* **78**, 4801.
Winstein, S., Klinedinst, P. E. and Robinson, G. C. (1961). *J. Am. chem. Soc.* **83**, 885.
Winstein, S. and Robinson, G. C. (1958). *J. Am. chem. Soc.* **80**, 169.
Winstein, S., Savedoff, L. G., Smith, S., Stevens, I. D. R. and Gall, J. S. (1960). *Tetrahedron Lett.* **9**, 24.
Winstein, S., Smith, S. and Darwish, D. (1959). *J. Am. chem. Soc.* **81**, 5511.
Winstein, S. and Trifan, D. (1952). *J. Am. chem. Soc.* **74**, 1154.
Wolfe, R. H. and Young, W. G. (1956). *Chem. Rev.* **56**, 753.
Zollinger, H. (1964). *Adv. Phys. Org. Chem.* **2**, 163.

Index